e-topia

"URBAN LIFE, JIM—BUT NOT AS WE KNOW IT"

e-topia

"URBAN LIFE, JIM—BUT NOT AS WE KNOW IT"

WILLIAM J. MITCHELL

THE MIT PRESS / CAMBRIDGE, MA / LONDON, ENGLAND

First MIT Press paperback edition, 2000

© 1999 MASSACHUSETTS INSTITUTE OF TECHNOLOGY

This book was set in Rotis Semi Serif and Bembo by The MIT Press and was printed and bound in the United States of America.

LIBRARY OF CONGRESS CATALOGING-IN-PUBLICATION DATA

Mitchell, William J. (William John), 1944–
 E-topia : "Urban life, Jim—but not as we know it" / William J. Mitchell.
 p. cm.
Includes bibliographical references and index.
ISBN 0-262-13355-5 (hc : alk. paper), 0-262-63205-5(pb)
1. Telecommunication—Social aspects.   2. Computer networks—
Social aspects.   3. Information superhighway—Social aspects.
4. Cities and towns.   I. Title.
HE7631.M58   1999
303.48'33—dc21   99-20670
                                                          CIP

CONTENTS

For Emily and Jane

e-topia

"URBAN LIFE, JIM—BUT NOT AS WE KNOW IT"

PROLOGUE: URBAN REQUIEM

Marshall McLuhan, 1967: "The city no longer exists, except as a cultural ghost for tourists."[1]

Yes, yes, I know; it's a familiar trope—death of God, death of the subject, death of the author, death of the drive-in, end of history, exhaustion of science, whatever. But he turned out to be right—though a few decades ahead of his time, as usual.

It's finally flatlining. The city—as understood by urban theorists from Plato and Aristotle to Lewis Mumford and Jane Jacobs—can no longer hang together and function as it could in earlier times.[2] It's due to bits; they've done it in. Traditional urban patterns cannot coexist with cyberspace.

But long live the new, network-mediated metropolis of the digital electronic era.

■ **The First Mourner's Eulogy**

DOA at Y2K! Whatever happened to the city as we know it?

I'll tell the tale.

Long ago, there was a desert village with a well at its center. The houses clustered within the distance that a jar of water could comfortably be carried. In the cool of the evening the people came to the well to collect the next day's supply of water, and they lingered there to exchange gossip and conduct business with one another. The well supplied a scarce and necessary resource, and in doing so also became the social center—the gathering place that held the community together.

Then the piped water supply came. Who could deny the practical advantages? It was more convenient, and kids no longer got cholera. Population grew, and the village expanded into a large town, since houses could be supplied with water wherever the pipes could run.

Dwellings no longer had to concentrate themselves in the old center. And the people ceased to gather at the well, since they could get water anytime, anyplace. So the space around the wellhead lost its ancient communal function, and the people invented some new, more up-to-date and specialized sites for socializing—a piazza, a market, and a cafe.

History replays—this time because the information supply system has changed. Once, we had to go places to do things; we went to work, we went home, we went to the theater, we went to conferences, we went to the local bar—and sometimes we just went out. Now we have pipes for bits—high-capacity digital networks to deliver information whenever and wherever we want it. These allow us to do many things without going anywhere. So the old gathering places no longer attract us. Organizations fragment and disperse. Urban centers cannot hold. Public life seems to be slipping away.

Take something as simple but telling as a day at the races. Before telecommunications, this involved traveling to the racecourse, mixing with punters in the stands, placing your bets with bookies on the rails, watching the horses with your own eyes, and settling your wagers face to face. Then, when radio and the telephone came along, races were broadcast, off-track betting (both legal and illegal) flourished, and on race days you could hang out at different places—at pubs and betting shops. Now, the ever-entrepreneurial Hong Kong Jockey Club has reconfigured the system once again by introducing handheld, electronic, networked devices that allow you to place your bets from anywhere in the city, at any time of day. You just need a telephone jack or a wireless connection to log in, and the system settles your accounts automatically. It is extraordinarily efficient, but it also eliminates occasions that going to the track had provided for making contacts, socializing, building trust, and doing deals.

Once again, we need to innovate—to reinvent public places, towns, and cities for the twenty-first century.

■ The Second Mourner's Eulogy

And that's not all. Digital communication also remakes the traditional rhythms of daily life.

Not so long ago, a family of the North lived in a fine clapboard house. There was a chimney at the heart of it, and to keep in the warmth the walls formed a simple surrounding box. In the winter, family members gathered round the fireplace—which was the only source of heat and light. Here, the children studied, the parents exchanged news of the day, and Grandma worked at her embroidery. The hearth held the extended family together.

Then pipes for delivering energy were put in—electrical wiring and central heating ducts. Family members could be warm and have light to read by everywhere. The fire was no longer kindled, except as a kind of nostalgic entertainment on festive occasions. The kids withdrew to their rooms to do their homework and listen to their stereos. The parents began to work different shifts, and would leave testy notes for each other on the refrigerator door. Grandma got bored and cranky, and soon moved out to an air-conditioned nursing home near Phoenix where she could play bingo with her similarly sidelined cronies. The fireside circle could no longer serve as social glue.

Informatization is following hard on the heels of electrification, with social consequences that are at least as profound. As the engineers figure out the technology, and the venture capitalists keep the IPOs popping, tiny telecommunications and information-processing devices are becoming as commonplace as lightbulbs and electric motors. You can call just about anyone, anywhere in the world, at any moment, from your digital cell phone. You can have twenty-four-hour news delivered digitally, by satellite, to your hotel room TV. You can pick up your email, whenever you want it, at any telephone jack. You can get cash at any ATM, any time. Your domestic appliances have embedded processors, and will increasingly require network connections as well as electrical and plumbing hookups. Your car is crammed with sophisticated electronics, and the guy who fixes it needs a computer as well as a wrench. The early industrial age of dumb devices is over; things now tirelessly, twenty-four/seven, think and link.

Today, ubiquitously present telecommunications networks, smart machines, and intelligent buildings combine with water supply and waste removal, energy distribution, and transportation systems to create a wherever, whenever, globally interlinked world. The old social fabric—tied together by enforced commonalties of location and schedule—no longer coheres.

What shall replace it?

■ The Third Mourner's Eulogy

Once, the Buddha sat under a bo tree. Disciples gathered in the shade and listened to his voice. To learn, they had to come within earshot. And in that place they formed their community of believers.

There was no other way.

Then his words were written down. First, the laboriously hand-written holy books were kept in monastery libraries, where the faithful could come to read; long after he was dead, they could travel to these book-centered communities as their predecessors had once come to the bo tree. Later, the books were printed, and the word could be delivered worldwide, to anyone who sought it. It was the same with other faiths. Though journeying to the holy sites survived as a spiritual exercise, and places like Santiago de Compostela and Mecca retained their magnetism, pilgrimage lost its more directly practical function.

As printed books proliferated and literacy spread, elaborate systems for storage and distribution of texts—both sacred and secular—sprang up everywhere. These took many scales and forms; there were national libraries, monastery libraries, university libraries, subscription libraries, municipal free libraries, suburban branch libraries, Carnegie libraries, Christian Science reading rooms, book-lined studies, book clubs, and bookmobiles. Main Streets had their bookstores and newsstands. Waiting rooms had their stacks of dog-eared magazines. Businesses depended on orders, ledgers, and invoices. Offices overflowed with files, briefcases were stuffed with paperwork, and even pockets held notes, cards, photographs, and paper money. Mail systems moved all this ink-on-cellulose around. Information was mobilized, and access to it was decentralized.

Today, text and images float free even from paper, and are pumped around at amazing speed through computer networks. We have online databases, Web sites, FAQs, and search engines. Email is rapidly replacing snail mail. In our technological age, seekers of enlightenment no longer need to embark on wearisome trips to distant sources of information. They don't even have to go to their local libraries. Bookstores, newsstands, magazine racks, theaters, temples, and churches—even bo trees—have their virtual equivalents. Students surf into electronic encyclopedias. Professors put their lecture notes up on the Web. Retailers put catalogs and order forms online. Stock markets speed quotes electronically to the screens of traders.

Mindwork no longer demands legwork. Commerce isn't impeded by distance. Community doesn't have to depend on propinquity. Links among people are formed in hitherto unimaginable ways.

Perhaps this new social glue can be turned to our advantage. Maybe homes and workplaces, transportation systems, and the emerging digital telecommunications infrastructure can be reconnected and reorganized to create fresh urban relationships, processes, and patterns that have the social and cultural qualities we seek for the twenty-first century. Maybe there's another way—a graceful, sustainable, and liberating one.

Two tentative cheers for the global village!

■ Mondo 2K+

How will it all play out? And what is to be done?

The buildings, neighborhoods, towns, and cities that emerge from the unfolding digital revolution will retain much of what is familiar to us today. But superimposed on the residues and remnants of the past, like the newer neural structures over that old lizard brain of ours, will be a global construction of high-speed telecommunications links, smart places, and increasingly indispensable software.

This latest layer will shift the functions and values of existing urban elements, and radically remake their relationships. The resulting new urban tissues will be characterized by live/work dwellings, twenty-four-hour neighborhoods, loose-knit, far-flung configurations of electronically mediated meeting places, flexible, decentralized production, marketing and distribution systems, and electronically summoned and delivered services. This will redefine the intellectual and professional agenda of architects, urban designers, and others who care about the spaces and places in which we spend our daily lives.

■ Doing Your Bit

This new agenda separates itself naturally into several distinct levels—
the subjects of following chapters. We must put in the necessary digital telecommunications *infrastructure,* create innovative *smart places*
from electronic hardware as well as traditional architectural elements,
and develop the *software* that activates those places and makes them
useful. Finally, we must imagine the architectural, neighborhood,
urban, and regional *spatial configurations* that will be sustainable and
will make economic, social, and cultural sense in an electronically
interconnected and shrunken world—a world in which distance has
lost some of its old sting, but also much of its capacity to keep chal-
lenges and threats comfortably removed.

To pursue this agenda effectively, we must extend the definitions
of architecture and urban design to encompass virtual places as well
as physical ones, software as well as hardware, and interconnection by
means of telecommunications links as well as by physical adjacencies
and transportation systems. And we must recognize that the fundamental web of relationships among homes, workplaces, and sources
of everyday supplies and services—the essential bonds that hold cities
together—may now be formed in new and unorthodox ways.

It is, I suggest, a moment to reinvent urban design and development and to rethink the role of architecture. The payoffs are high,
and so are the risks. But we have no choice; we cannot realistically
opt out. We must learn to build e-topias—electronically serviced,
globally linked cities for the dawning *début de K.*

You say you want a revolution? You want digital technology to deliver new and improved cities? Well, you know, most of the things promised by the digerati just haven't been up there with liberty, equality, and fraternity.

Tiny digital cell phones? Status toys for overgrown boys. *HDTV?* Great eyeball, no doubt, but garbage on a bigger screen is still garbage. *Movies on demand?* Marginal social benefit at best. *Virtual reality video games?* Fun for a few minutes. *Your very own home page on the Web?* Electronic vanity publishing. *Push-served sports scores?* Please! Wired whiz-bang today, tired techno-yawner tomorrow.

So don't look here for more techno-triumphalist, macho-millennial prophecies of a glittering, go-ahead cyberfuture. But don't expect equally dogmatic and deterministic Chicken Little inversions of these visions either—reiterations of those now-familiar glum assertions that the digital revolution must inevitably reinscribe the nastier existing patterns of power and privilege, while trampling on treasured traditions as it does so.

■ Digiphiles versus Digiphobes

We know, by now, the tiresomely predictable ideological subtexts to these polar positions. From the government-butt-out right comes the view that digital technology *can* improve our lot, therefore it *will*— provided we don't mess with the market. From the policy-wonk left comes the rejoinder that the rich and powerful are always the first to benefit from new technologies, and that markets are no friend of the marginalized, so we need vigorous government intervention to guarantee that computers and telecommunications don't end up creating a yawning digital divide between haves and have-nots. And, of course, the neo-Luddites are firmly convinced that we all have far more to lose than to gain anyway, so we should just dig in and resist.

But the increasingly boring digiphiles and digiphobes, with their contending visions of utopia and dystopia, are myopically groping different extremities of the pachyderm. We will do far better to sidestep the well-known trap of naive technological determinism, to renounce the symmetrical forms of fatalism proposed by booster-technocrats and curmudgeonly techno-scoffers, and to begin, instead, by developing a broad, critical, action-oriented perspective on the technological, economic, social, and cultural reality of what's actually going on, all around us, right now.[1] Since new technological systems are complex social constructions, we must understand our emerging options, choose our ends carefully, and build well.[2] Our job is to design the future we want, not to predict its predetermined path.

■ After the (Digital) Revolution

Begin by looking around you. Your own eyes and the accumulating social science evidence should swiftly convince you—if you're not persuaded already—that the digital revolution cannot be dismissed as mere hype and hyperbole. This trumpeted technological transformation, which we're assured has been "whipping through our lives like a Bengali Typhoon," has actually been real enough.[3]

Somewhere around 1993, with the takeoff of the World Wide Web and the launch of *Wired* magazine, this geekerati-led, network-enabled, silicon-powered insurrection against the old order had its virtual 1789, October, May 4th, or . . . fill in your favorite. It became obvious to the observant that familiar regimes were being swept away by simultaneously unfolding, causally intertwined processes of technological innovation, capital mobilization, social reorganization, and cultural transformation.

As with those monster shakeups that have punctuated our past— the agricultural and urban revolutions following from the invention of the wheel and the plow, and the industrial revolution which emerged from Enlightenment science—the postrevolutionary social dynamics have gathered seemingly unstoppable momentum. They are rocking our institutions and roiling our surroundings. They are creating new opportunities and closing off some old ones. Their effects will not always be as advertised by the cheerleaders, they will not be wholly positive, and they will not be uniformly distributed, but they cannot be ignored.

To understand this particular transformation's trajectory, we must recognize that—like its big-time history-book predecessors—it is not really the product of a single dramatic event. Nor is it the consequence of some self-contained invention. It has resulted, instead, from the gradual convergence of several extended processes. Until recently, these were puttering along in parallel. But when they came together, it was like mixing the otherwise innocuous elements of nitroglycerin. Then the World Wide Web supplied the spark, and the result was an explosively exponential expansion—a Big Bang that's the beginning of something genuinely new.

Specifically, the crucial ingredients of the incendiary brew have been digital information *storage, transmission, networking*, and *processing* hardware, together with the associated *software* and *interface* capabilities.[4] Products and services based on these various technologies are now produced and distributed on a wide economic front—by the telephone, radio and television, cable TV, semiconductor, computer, consumer electronics, software, publishing, and entertainment industries—and these industries have become increasingly interlocked and interdependent. Information has become dematerialized and disembodied; it is now whizzing round the world at warp speed, and in cortex-crackling quantities, through computer networks. And this vast global process is just booting up.

■ Information, Infrastructure, and Opportunity

The broad outlines of our electronically mediated future—if not the details—are becoming clear. One way or another, depending upon the eventual outcomes of the technology races, business battles, and public policy debates of the millennium's.end, these disparate ingredients will eventually combine to produce a worldwide digital information infrastructure.[5] The potential benefits of this are so great, and the momentum for it is building so rapidly, that nothing will effectively stand in the way.

This emergent system will combine the comprehensive geographic coverage and sophisticated person-to-person and place-to-place connection capabilities that characterize the existing telephone system with the high-speed pipes and multimedia affordances of cable television. And it will add to the mix the virtually limitless storage capacity and processing power produced by the silicon chip. Prefixes describing all aspects of its capacity will continue to crank up from *kilo* to *mega* to *giga* to *tera*—even *peta* and beyond.[6]

Physically, it will be a complex construction of computational devices, copper wires, coaxial cables, fiber optics, wireless communications systems of various kinds, and communications satellites. Logically, it will be held together by widely accepted conventions and protocols with indigestibly acronymic names like TCP/IP, HTTP, FDDI, and ADSL. Economically, it will be the joint creation of innumerable, widely distributed businesses and public authorities with very different sorts of stakes in the system and diverse ways of making money from it. It is being created incrementally and messily through a complex ongoing process of technological innovation, new infrastructure construction, adaptive reuse of existing infrastructure, alliances and mergers among telecommunications providers, and reformulation of regulatory regimes.

Eventually, information of every kind will collect in a planetful of computers, and will be delivered wherever you want it through a single digital channel. Everyday objects—from wristwatches to wallboard—will become smarter and smarter, and will serve as our interfaces to the ubiquitous digital world. And paradoxically, wherever you happen to come in contact with this immense collective construction, it will seem to have the intimacy of underwear.

Instead of forming new relationships of people and agricultural production sites as in the agricultural revolution, or of people and machines as in the industrial revolution, this global digital network will reconstitute relationships of people and *information*. It will increasingly become the key to opportunity and development, and the enabler of new social constructions and urban patterns. Investment, jobs, and economic power seem certain to migrate to those neighborhoods, cities, regions, and nations that can quickly put the infrastructure in place and effectively exploit it.[7]

New Networks and Urban Transformation

As historically minded observers can scarcely fail to anticipate, this latest wave of urban infrastructural networking will play much the role that its predecessors did in earlier eras of technologically mediated metamorphosis—in the times of Roman roads and aqueducts, in the boom time of eighteenth-century shipping and waterways, in the heyday of the nineteenth-century railroad robber barons, and in the expansion years of the twentieth century's electricity grids and Interstates.[8] As canals and muscle power were to Amsterdam, Venice, and Suzhou, as tracks, ties, and steam trains were to the open spaces of the American West, as the tunnels of the Underground were to London, as the internal combustion engine and the concrete freeway were to the suburbs of southern California, and as electrification and air conditioners were to Phoenix, so the digital telecommunications system will be to the cities of the twenty-first century.[9]

Like their pipe-and-wire predecessors, however, digital telecommunications networks will not create entirely new urban patterns from the ground up; they will begin by morphing existing ones. Generally in the past, new urban networks have started by connecting existing activity nodes that had been made possible and sustained by earlier networks. (After all, what else *could* there be to connect?) Then, like parasites taking over their hosts, they have transformed the functioning of the systems on which they were superimposed, redistributed activities within these systems, and eventually extended them in unprecedented ways.

Thus the coming of the railroads transformed the existing settlement of Chicago into a pivotal national center as the West opened up; then road and air transportation remade it once again. In southern California, an extensive rail system initially linked together a system of small towns scattered through the valleys; then the freeway network reconnected them, allowed the spaces in between to develop, and eventually wove the pattern that we now know as the modern Los Angeles metropolitan region. And in the twenty-first century, new, high-speed, digital telecommunications infrastructure will refashion the urban patterns that emerged from nineteenth- and twentieth-century transportation, water supply and waste removal, electric power supply, and telephone networks.

You can already see this sort of transformation unfolding in the pleasant Indian city of Bangalore, for example. Bangalore initially grew, on an ancient foundation, as the capital of a princely Mysore state. Then, in the British era, it became a railway center. From the second half of the nineteenth century, its accessibility, pleasant climate, and green, attractive surroundings attracted administrative activities, industry, educational and research institutions, and eventually a large population of well-educated professionals. By the 1990s, it had a new infrastructure of satellite earth stations, microwave links, and software parks, and through this it had become a thriving center of the software export industry. Bangalore software enterprises could compete effectively on the world market by employing high-speed electronic links to import intellectual raw materials, export finished software products, and interact with their clients, while tapping into a skilled but relatively inexpensive local talent pool.

It's an old script replayed with new actors. Silicon is the new steel, and the Internet is the new railroad.

■ The Big Pipes

New urban infrastructures tend to be Viagra versions of older, tireder predecessors that cannot quite do the job any more. Their enhanced potency makes a qualitative difference. When piped systems replace wells you get a greater flow of water and you can take long, hot showers. When freeways supplant dirt tracks you can live in the suburbs and drive every day to work. And when high-speed, digital telecommunications systems succeed the telegraph and the telephone, you get socially significant changes in everyday interactions. It turns out that the more bits per second you can push through a communications channel, the more complex and sophisticated the interchanges and transactions that can take place over it.

This was evident right from the beginning of electronic telecommunications. The telegraph carried single-toned dots and dashes over an iron wire, it was excruciatingly slow and very expensive, and its limitations left us the word "telegraphic" to describe the terse and abbreviated style of textual discourse that it engendered. The range of frequencies required for speech transmission demanded greater bandwidth, so the telephone system used copper wire to provide it.[10]

At the low end of modern digital telecommunications, there is the world of one kilobit per second communications—as provided by early modems and by the French Minitel system. At this rate (or less) it is feasible to exchange short text messages. This suffices for limited social, educational, and commercial interaction via electronic mail—for setting up meetings, for routine transactions such as placing orders, checking inventories and account balances, and paying bills, and for creating elementary, text-based forms of virtual public space such as bulletin boards, Usenet newsgroups, and MUDs and MOOs.

Jump up an order of magnitude or two; at tens to hundreds of kilobits per second (as provided, for instance, by a 28.8 kilobits per second modem or a 128 kilobits per second ISDN connection), large text files and high-quality color graphics can be moved around with adequate speed. This level of connection was very widely available by the mid-1990s. Together with the high-speed backbone of the Internet (which was designed to operate at 45–155 megabits per second), it allowed the World Wide Web to grow at a remarkable rate. By providing an online equivalent to printed books, magazines, and catalogs, the Web opened the way to online publishing, advertising, and retailing on a significant scale. Virtual bookstores and newsstands began to compete with physical ones, and virtual malls and campuses began to appear. But the graphics of the early Web were mostly two-dimensional, and navigation was just pointing and clicking.

Now move to the megabit range; at rates of megabits per second to tens of megabits per second, good audio and video are possible, graphics can become very sophisticated, and elaborate, three-dimensional, shared virtual worlds can be created. These transfer rates have long been provided to homes by cable television networks, but only one-way—from the provider to the consumer—rather than symmetrically. They have also been provided by the local-area networks (LANs) and Internet connections of universities and large corporations; these have typically delivered about 10 megabits per second to the desktop, with faster systems running at 100 megabits per second. Over longer distances, lines leased from telecommunications providers have supplied T1 (1.54 megabits per second) and T3 (45 megabits per second) service.

At megabit and gigabit rates, expressive subtleties—tones of voice, body language, and so on—need not be filtered out, as they usually are in lower-bandwidth telecommunications. Furthermore, a great deal of useful context can be provided in the form of video backgrounds, shared access to work tools and materials, and shared virtual worlds—much as an architectural setting like an office or classroom provides an appropriate context for the activities that it accommodates. Thus telepresence can begin to compete effectively with bodily presence in situations—such as negotiating a contract, discussing a design proposal, or conducting a medical examination—where nuance and context are critical.

When these high rates are reached, networks actually run at speeds comparable to the processors and internal buses of computers. Consequently, computers begin to lose their discrete spatial identities; any scattered collection of interconnected processors and memory devices may become the functional equivalent of a PC in a box. As a slogan popularized (a bit before its time) by Sun Microsystems puts it, the network *is* the computer. This is where we are going to end up.

■ Connected to the Backbone

This all-encompassing digital system will create new linkages *between* cities and *within* cities, and its intercity and intracity components should carefully be distinguished. To begin with, there are significant technical and cost differences among local-area, metropolitan-area, and long-distance networks. But more importantly, they differ in their implications for urban life and form.

Long-distance, intercity linkages are formed by interconnecting major switching centers with high-capacity fiber-optic cables, microwave links, or satellite links to create digital telecommunications *backbones*. The switching centers are usually known as *POPs*—points of presence. If they are on backbones that run at gigabit rates, they are *gigaPOPs*. And large centers built around satellite earth stations have sometimes been promoted as *teleports*.[11]

Whatever form they take, these switching nodes on backbones—like seaports and airports before them—serve as points of connection to a wider world and potential generators of economic activity in their surrounding regions. It will be economically vital to have an efficient POP on the high-speed backbone in your vicinity. It will be an increasingly important competitive advantage if you have one and your business rivals do not. And equity considerations will motivate public policies that encourage wide and even distribution of POPs.

This pattern is clearest in developing countries, where introduction of a POP into a hitherto unserved region can make a sudden, vivid difference. During the 1980s and 1990s, for example, the government of India invested in high-speed satellite earth stations at Bangalore, Hyderabad, Pune, Noida, Bhubaneshwar, Thiruvananthapuram, and Chandigarh. These provided twenty-four-hour international connectivity to nearby software parks containing workspace for software enterprises, and thus became the focal points of the thriving software export industry.[12] (In less than a decade, India became the world's largest exporter of teleservices, and the second-largest exporter of software.)[13] Since there was little high-speed terrestrial infrastructure, the effects were mostly felt in the immediate vicinity—at most, over the twenty-to-thirty-kilometer radius typically reachable by microwave link from a transmission tower. In effect, they created digital oases.

In developed countries, the digital revolution has unfolded in a context of established telephone and cable telephone infrastructure that could be adapted to carry digital data, and this has made the situation more complex. You can get digital connection almost anywhere—typically from many competing providers—but speeds, costs, and levels of reliability vary widely.

■ New Global Interdependencies

The most dramatic general effect of this long-distance digital telecommunications infrastructure is to create new kinds of interdependencies among scattered regions and settlements. For example, businesses have discovered that low-cost, high-quality voice and video connections enable delivery of certain customer services from great distances; being in the right time zone, speaking the right language, having the right software, and being competitive in a global labor market can become more important than being in the same metropolitan area.

Thus telephone and video call centers in Sydney can serve customers who want to make airline reservations in Hong Kong. Similarly, stenographers in Hyderabad can transcribe dictation from doctors in Chicago (exploiting the time zone difference to provide overnight service), draftsmen in Manila can produce CAD documents for London architectural and engineering firms, and very-low-wage workers in Africa can watch video monitors connected to security cameras in New York.

Such interdependence is not, of course, an unprecedented phenomenon. Neighboring cities have often traded with one another, and in the past new infrastructures have created expanding systems of economically, politically, and culturally interdependent settlements. In the United States, for instance, the interurban network that holds the nation together began as a line of port cities along the Atlantic coast, then reached westward to the Mississippi as new cities developed along inland waterways, and eventually extended coast to coast in the era of the railroads and the telegraph.[14] Even economic and cultural globalization long predates the computer and the communications satellite, as many observers have noted.

The point, though, is that digital telecommunications infrastructure greatly increases the *density* of linkages within systems of cities, and can spread these systems worldwide. The electronic interconnection of currency traders to form a high-speed global trading system provides the most dramatic illustration of this, but it is really just an early straw in the digital wind.[15] There is much more in the works.

■ From POP to Your Door

In general, when local networks of any kind are created and linked to long-distance networks, they diffuse the benefits of distant connection among the inhabitants of their service areas. Connecting a local water supply system to an aqueduct brings water from a distant source directly into homes. Linking local roads to the Interstate allows small-town businesses to benefit from passing traffic. (Conversely, getting bypassed by the Interstate can be a disaster for them.) And hooking local digital networks to POPs on high-speed, long-distance backbones puts a populace in direct touch with the world.

Creating the local loops from POPs to homes and businesses is an expensive and time-consuming task, though, since there are so many of them to provide, and since provision typically involves digging up streets; providers face what they often call "first mile" and "last mile" problems.[16] How do potential customers link their sites to the nearest POP? How do providers get from their POPs to all those potential customers out there? Who pays for these local loops? And how do costs get recovered? Providers attempt to solve these problems not only by putting new local infrastructure in place, but also by adapting existing telephone, cable television, and even electric power lines to the new task of digital telecommunications.

For individuals, these POP-to-doorstep connections offer a partial escape from the old need to choose between intimate, supportive, yet often-constricting local communities on the one hand and the opportunities that seem inseparable from the anonymity and alienation of the big city on the other—*Gemeinschaft* versus *Gesellschaft*, in the famous formulation of Ferdinand Tönnies.[17] It was a geographic choice: one sort of place or the other. In an era of interlinked digital networks, though, you can live in a small community while maintaining effective connections to a far wider and more diverse world—virtual *Gesellschaft*, as we might term it, without tongue too far in cheek. Conversely, you can emigrate to a far city, or be continually on the road, yet maintain close contact with your hometown and your family—electronically sustained *Gemeinschaft*.

It's not all good news, however. Those very same liberating connections create competition between local and distant suppliers of goods and services, and can shake a local community's economic and cultural foundations. Recall that local wells fall into disuse when the piped water supply comes. When customers begin to take superhighways to regional malls, the local stores lose out. Local radio and television shows must contend with network offerings that go out to much wider audiences, and so can afford bigger stars and fancier production. And when local digital networks hook up to the backbone, many of the familiar protections of isolation and transportation cost disappear, and distant competitors can take vigorous advantage of the openings that result.

■ **The Network City Extended**

Intraurban digital networking furthers the long evolution of human settlements from loose collections of more or less independent dwellings to highly integrated, networked cities in which multiple infrastructures of tracks, pipes, and wires deliver centrally supplied services to buildings and carry away waste.

The incipient networked city is clearly visible in the ruins of Pompeii, with its hillside civic reservoir, network of lead water-supply pipes running down through the town, and gravity-fed waste-water drainage system. In the aftermath of the industrial revolution, cities greatly elaborated their networks by improving streets to handle greater traffic volumes, adding streetcar and rail transportation systems to meet the demands of larger and more widely distributed populations, constructing municipal water supply and sewage systems to improve sanitation, creating gas and electric utilities to distribute energy, and eventually adding local telephone networks for communication.[18] Digital data distribution systems will soon become as ubiquitous within cities as electrical and telephone networks, they will carry many different kinds of information, and they will ultimately (if not immediately) provide high capacity at low cost.

From the viewpoint of businesses with offerings that can be ordered or distributed electronically, the new intraurban digital networks create easily reachable consumer markets.[19] Thus they are crucial to news and entertainment companies, publishers, banks, and online retailers. Not surprisingly, then, they have quickly become fierce competitive battlegrounds and subjects of study in the trendier business schools. At the same time, they create a powerful alternative to intermediate distribution sites such as local newsstands, video stores, movie theaters, and branch banks—and may, indeed, threaten the very existence of these established neighborhood elements.

Seen from the differing perspective of local educational and cultural organizations, government agencies, community activists, and politicians, these same intraurban networks potentially provide an updated version of the agoras and forums of the past, a new means of strengthening interactions within communities, and a mechanism for discussion and organization. So they have encouraged dreams of a reinvigorated Jeffersonian democracy, spawned a grassroots "community networks" movement, and supported the emergence of popular online meeting places such as the San Francisco Bay Area's Well and New York's Echo.[20]

■ The End of Rural Isolation?

Digital networks can, however, extend much further than the networks of the past—so much so that they challenge long-established distinctions between urban and rural areas.

Once these distinctions seemed pretty clear. Many old depictions of urban scenes, such as Pietro and Ambrogio Lorenzetti's famous *Good and Bad Government* panels in Siena's Palazzo Pubblico, have vividly shown how the city's limits were defined by its walls. Outside was the countryside, with its rustics, recluses, and assorted inconveniences and dangers. Urban expansion was accomplished, if necessary, by enclosing additional area; you can clearly trace the increments of growth in the street patterns of many old European cities.

Even in ancient times, though, it was not always quite so simple. Athens, for example, was largely a community of independent farmers who lived outside the walls, and came to town from time to time. Meeting places and other communal facilities were concentrated at the center, and a network of paths and roads extended out into the hinterland.

The far more elaborately networked cities of the nineteenth and twentieth centuries dispensed entirely with walls, and characteristically grew by extending their infrastructures. Being beyond the metropolitan limits came to mean being past the reach of the trolley car lines, the water supply system, and the sewers. These networks tended to thin out gradually, rather than disappear suddenly, with increasing distance from urban centers.

It subsequently turned out that wired infrastructure—the electricity grid and the telephone system—could be extended into the closer and more densely populated rural areas with particular ease. In the twentieth century, then, rural electrification and telephone systems have done a great deal to improve the conditions of life outside the city limits.

Digital telecommunications infrastructure is now beginning to follow the old electric and telephone wires, and in some cases to piggyback on the existing copper. (Less obviously, it can even make use of existing railway signal lines and wire fences.) And even the most minimal rural telecommunications infrastructure, strategically deployed, can have dramatic social and economic effects. India, for example, has pursued a successful program of providing rural telephone service through village-to-village lines, small, highly robust switches, and public telephones with attendants who can provide assistance to those unfamiliar with the technology; it is a natural next step to extend these facilities to fax and to public Internet access. Vastly improved access to emergency services is the immediate result. Longer term, this new linkage promises to change rural economic life by providing farmers with direct access to distant buyers for their produce, and to transform rural education by providing minimal but effective access to the resources of the World Wide Web.

But even more importantly, wireless systems—both terrestrial and satellite—are now providing an extraordinarily effective new way to reach rural inhabitants.[21] Microwave links and wireless cellular systems can traverse large stretches of rough terrain simply by means of some strategically placed transceiver towers. During the 1980s and 1990s, for example, the Australian telecommunications provider Telstra constructed an extensive system of solar-powered microwave repeater towers across the empty expanses of the Outback. These landmarks pop up along the roads at intervals of about fifty kilometers—providing travelers with a new measure of distance.

Satellite telecommunications systems are not affected by terrain at all, and can deliver services even more economically to areas with very low population densities and teledensities (telephone lines per hundred residents).[22] Older geosynchronous satellite systems had large but limited service footprints, and mostly focused their capacity on densely populated areas. But newer LEO (low earth orbit) systems, such as Iridium and Teledesic, uniformly blanket the earth.

As rural telecommunications infrastructure begins to deliver increasingly sophisticated educational, medical, and other vital services, then, the old distinctions between city and countryside, and between center and periphery, are becoming fuzzier and fuzzier. This continues a transformation that began long ago. In one of their most famous passages, Marx and Engels observed that the growth of great industrial cities had "rescued a considerable part of the population from the idiocy of rural life."[23] Today, the digital revolution is completing the job.

■ Residual Wireless Backblocks

Nonetheless, telecommunications capabilities will remain scarcer in the far-away, less-developed wireless backblocks—way out where the tumbleweeds blow, and on Micronesian coral specks—than they are in sophisticated urban areas. And this will yield characteristically different usage patterns.

Sometimes rural dwellers need information in a hurry. If they need answers to emergency medical queries, for example, they need them right away. And rural development, disaster relief, and rehabilitation workers often have critical, time-dependent information requirements. In these cases, short-term access to the most advanced telecommunications facilities is what's needed. So grabbing a satellite link for a while—even though it is comparatively expensive—may make sense.

But in many other cases, less dramatic reduction in times taken to obtain answers to queries—from months or weeks to days or hours—suffices to make a huge difference in the quality of medical care, education, and other vital services. So there is growing interest in using small amounts of telecommunications capacity to provide very inexpensive "real-time-enough" email messaging services to poor and isolated rural areas. A system called Fidonet effectively pioneered this strategy by employing off-peak dial-up links and batched transmission of email messages.

Now, such low-end, low-cost services can begin to take advantage of the fact that LEO communications satellites are doing almost nothing, and so have spare capacity, when they are passing over sparsely populated areas. As Nicholas Negroponte has put it, "With LEOs, you have to cover the whole world in order for any single part of it to work—rural and remote access, in a sense, comes free."[24]

Even with such improvements, though, residents of the wireless rural backblocks will continue to suffer from some disadvantages, due to an inherent asymmetry in airborne telecommunications; it is usually much cheaper and easier to build a big, central transmitter that blasts information out over a wide area than it is to build numerous distributed transmitters that send information back. Thus it is easier to provide high-speed downlink service to rural areas—particularly from satellites—than it is to provide equivalent uplinks. So rural residents tend to get broadcast and Web downlink service (together, typically, with low-capacity back channels) long before they get the capacity to pump large amounts of information back out to the rest of the world.

■ **Public and Private**

Much of this emergent telecommunications infrastructure—local and long-distance, urban and rural—is being created and maintained by organizations that are in the bit-hauling business. By itself, though, bit hauling is not a terribly attractive kind of work for private-sector organizations to pursue; digital telecommunications capacity is a low-cost commodity, generating low profit margins, so many of the players attempt to do better by adding value to flows of bits—for example, by creating and distributing entertainment or strategically inserting advertising. The structure that results is a large-scale, widely available, heterogeneously used utility—much like the public road system. Hence the wearyingly overused "information superhighway" metaphor.

But there are numerous private networks as well. Some of these operate within buildings and campuses, like internal plumbing systems. Some are highly specialized EDI (electronic data interchange) networks linking businesses such as banks to one another. And some are private long-distance networks maintained by large, far-flung organizations and operating over lines leased from telecommunications providers.

Some of these private networks operate under specialized protocols, but increasingly many employ the same ones as the public Internet and World Wide Web, and make use of the same software. These have become known, in a triumph of techie prefix-mongering, as intranets. Symmetrically, networks used to create an organization's public presence may be known as extranets.

■ **Behind the Firewalls and Filters**

Where security is important, intranets and other private networks attempt to preserve their privacy through physical isolation and careful control of access points. Like fortresses of old, they have few connections to the outside world, and those connections are designed to allow very close supervision of everything that comes in and goes out. But instead of fortified gates and sentry posts, the connections between private intranets and the public Internet are formed by specially programmed computers that serve as electronic "gatekeepers." These ever-vigilant sentry devices determine when outsiders may have access, when insiders may make outside connections, and what sorts of information may flow back and forth. In doing so, they establish a clear distinction between the territory which is "inside the firewall" and its external environment.

The idea that information freely flows everywhere in a digitally networked world is, therefore, a wishful libertarian myth—or, if you are more worried about maintaining some control over access to certain information, a needlessly dark dystopia. Parents, teachers, employers, and governments can all create closely controlled online environments by isolating them behind tightly supervised connections to the public networks, and by defining internal rules and norms.[25] These controlled zones can range in scale from individual computers to entire nationwide networks.

The outcomes are complex. Ubiquitous interconnection does not mean the end of controllable territory, or elimination of distinctions between public and private turf, but it does force us to rethink and reinvent these essential constructs in a new context. The emerging system of boundaries and control points in cyberspace is less visible than the familiar frontiers, walls, gates, and doorways of the physical world, but it is no less real and politically potent.[26]

■ The Task Ahead

These effects of worldwide digital telecommunications infrastructure are powerful and sweeping, but it obscures the issue to claim—as some cyber-smitten hypesters hyperbolically have—that they will yield the death of distance, the end of space, and the virtualization of just about everything. (All that is solid melts, in this hot air.) It is more useful and illuminating, instead, to recognize that the resulting new linkages provide us with a radical new means of producing and organizing inhabited space, and of appropriating it for our multifarious human purposes.[27]

We all, therefore, have an immediate and vital interest in this mother of all networks, and in the social, economic, policy, and design questions that it raises. What new benefits might it bring, and what are they worth to us? How will it get constructed and paid for? How will it interact with existing urban patterns? Who will control it? Who will get access, and when? How might we balance incentives for telecommunications entrepreneurs and investors with policies that assure equity of access? What social and cultural qualities do we want this new mediator of our everyday lives to have?

The time and the fashion for breathless, the-world-is-new, anything-is-possible rhetoric have passed. And it turns out that we face neither millennium-any-day-now nor its mirror image—apocalypse-real-soon. Instead, we have been presented with the messy, difficult, long-term task of designing and building for our future—and making some crucial social choices as we do so—under permanently changed, postrevolutionary conditions.[28]

All networks produce privileged places at their junctions and access points.

There are fertile oases where irrigation networks pump out water—nowhere clearer than in the dramatic green circles created by center-point systems on the plains of the American West. Thriving businesses have developed around railway junctions, highway off-ramps, seaports on shipping routes, and air transportation hubs. In the nineteenth century, remote desert settlements like Alice Springs and Darwin were established as telegraph towns. And today, there are *smart places* where the bits flow abundantly and the physical and digital worlds overlap, at points where we plug into the digital telecommunications infrastructure.

From an architect's viewpoint, these electronically mediated places are not uniform, dimensionless nodes, as they rather misleadingly appear on the abstract network diagrams made by telecommunications engineers. Nor are they just plastic boxes stuffed with electronics. In fact, they have spatial extension, they engage our bodies, they are situated in particular physical contexts, and their spatial and material configurations matter. They are inhabited, used, and controlled by particular groups of people. They have their local customs and cultures. They range in scales and characters from the intimate and private to the globally public. And they are not just interfaces; we are beginning to live our lives *in* them.[1]

They have not only IP addresses but also street addresses. They provide not just electronic linkages to other smart places, but also doors and windows to physically adjacent spaces. Thus they are simultaneously embedded in and sustained by systems of physical and material circulation, visual and acoustic communication, and remote interconnection. By virtue of all these linkages, working in concert, they are starting to construct new contexts for our daily activities.

You can think of them as sites where two otherwise distinct domains—meatspace and cyberspace, as *Neuromancer* so vividly and provocatively troped it, or maybe biomass and infomass—are intersected, in some effective combination, to support some particular human activity.[2] They are places—as we shall see—where physical actions invoke computational processes, and where computational processes manifest themselves physically. The best of them will have the felicitous qualities that we have traditionally valued in our physical surroundings together with startling new affordances provided by widely available, inexpensive, electronic intelligence and telecommunication.

■ Proscenium and Display

In the seventeenth century, baroque theater architects confronted the similar task of bringing together the space of dramatic action and the space of the audience, and they accomplished the combination by means of the proscenium. It was a brilliant architectural invention. At the Teatro Farnese in Parma, Giovanni Battista Aleotti created a rectangular wooden structure with a stage at one end, seats at the other, and an elaborately framed opening with a stage curtain in between. This established the possibility of lighting the stage, dimming the auditorium, and giving audience members the convincing illusion that they were alone in the darkness spying on the characters through a virtual "fourth wall."

Within your living room, your television set creates a strikingly similar relationship by directly appropriating this idea. You sit on the audience side of a framed phosphorescent screen—sometimes even in darkness—and peer into an illuminated scene. Even the conventions of set design for television dramas echo those of the proscenium stage—and, in fact, the home viewer may get much the same perspective on the action as a studio audience.

On your desktop, your personal computer—an ungainly, immature, Frankenstein's-monster combination of television, typewriter, and disk player that will soon look as silly as biplanes and Model Ts—extends this tradition to yet another context. In the early days of PCs, you just saw scrolling text through the rectangular aperture, and the theatrical roots of the configuration were obscured. Then two-dimensional graphic desktops, with frontally rendered objects as in an Egyptian painting, became commonplace. Finally, as three-dimensional computer graphics became increasingly feasible, as online chat spaces with perspectival scenery and avatar-actors became popular,[3] and as digital video began to erase the distinction between PCs and TVs, the screen clearly became a proscenium again—a hole cut through the membrane separating the space of our bodies and our buildings from cyberspace.

You could *look* at screenspace, and you could get information from it, but you could not *enter* it. Paul Saffo vividly observed, "Two parallel universes currently exist—an everyday analog universe that we inhabit, and a newer digital universe created by humans, but inhabited by digital machines. We visit this digital world by peering through the portholes of our computer screens, and we manipulate it with keyboard and mouse much as a nuclear technician works with radioactive materials via glovebox and manipulator arms. Our machines manipulate the digital world directly, but are rarely aware of the analog world that surrounds their cyberspace."[4]

So the technology of computer graphic display was new but the architectural thinking wasn't. It was back to the baroque; Aleotti would immediately have recognized this well-worn setup.

■ Screenspace: S, M, L, and XXL

If this wasn't novel, though, it was certainly effective. With the emergence of the PC, the growth of networks, and ongoing advances in display technology, countless millions of glowing glass rectangles scattered throughout the world have served to construct an increasingly intricate interweaving of cyberspace and architecture. And it turns out that Godzilla was right; size matters—a lot. So does positioning relative to our bodies.

At the smallest scale, for example, the screens of wristwatch and shirt pocket devices create portable, personal connections—smart space wherever you happen to be. The slightly larger screens of laptop devices allow a kind of electronic camping out; you can choose a spot—temporary office, hotel room, airplane, park bench, cafe table— and set yourself up to work there.

In all these cases, the screen faces us, and we cherish the resulting privacy. (If you're like me, you choose the window seat on an airplane when you want to use your laptop, so that you don't have people looking over your shoulder.) But if you turn the screen outward, instead, it begins to function as a powerful means of self-representation; projects by the artist Krzysztof Wodiczko, such as *Alien Staff* and *Porte-Parole,* have explored this possibility—with due attention to its neo-Brechtian dimensions.[5]

At furniture and appliance scale, desktop PCs in offices, digital TVs in living rooms, point-of-sale devices in shops, and ATM machines in bank vestibules define the uses and characters of rooms. They are both part of the equipment and part of the decor. And they have mostly been assimilated to well-established models of interior decoration. Thus in the Seattle mansion of Bill Gates, there are both traditional windows looking out to Lake Washington and electronic windows which—as in a latter-day Versailles—provide this apparently irony-challenged cyberking with sweeping vistas out into the digital domain that he dominates.[6] With sly architectural wit, by contrast, Robert Venturi has turned the tables on such Star Trek futurism by assimilating display screens to the classical tradition of architectural decoration; in projects such as the rehabilitation of Harvard's Memorial Hall, he has treated LED screens as dynamic friezes, inscriptions, and murals.[7]

At this intermediate scale, screens typically function as electronic protagonists in social interactions. For example, a PC, ATM, or desktop videoconferencing screen sets up a one-to-one person-machine dialogue; indeed, the whole field of PC interface design presumes an individual user oriented nose-to-screen. At the airline ticket counter, the screen sits *between* the customer and the ticket clerk; it injects necessary information into the interaction between customer and clerk, and—because it faces only one way—it privileges the clerk. By contrast, a television screen in a living room or sports bar serves as a shared point of reference, and so establishes very different grounds for conversation and interaction—including, of course, arguments about who controls the remote! In classrooms and conference rooms, video projection screens now replace blackboards as sites of demonstration; the speaker controls and the audience watches.

At wall size, electronically animated screens can change the perceived character of the space itself. Video-projected screens with life-size human figures can, for instance, create the illusion that widely separated rooms have suddenly been jammed together and the dividing surface made transparent. It is dramatically effective—if rather unfortunately reminiscent of those prison visiting rooms where interaction takes place only through a glass screen.

In the 1980s, researchers at the Xerox Palo Alto Research Center experimented with virtually juxtaposing conference rooms and workspaces in this fashion. A little later, in his Clearboard system, Hiroshi Ishii made elegant use of the idea to create "transparent" drawing boards for remote design collaboration; you saw your collaborator "through" what appeared to be a double-sided drawing surface.[8] More recently, IBM has concocted "virtual dining" suites containing tables divided down the middle by back-projection screens on which you see life-sized video images of diners seated around your table's far-displaced opposite half. And Bruce Blumberg's ALIVE system presents itself as a large "magic mirror" in which full-scale, live video images of inhabitants interact with computer-generated "pets" and other animated elements.[9]

Finally, at urban scale—as in Times Square, Tokyo's Ginza, and at ballparks everywhere—giant electronic screens function like animated billboards, and can be employed to address vast crowds. If you don't mind replacing a lot of lightbulbs, this strategy can be pushed to a gigaglitz extreme; in Las Vegas, a 1,400-foot, 211-million-light, 54,000-watt-sound, programmed display screen has been used to create an entire new roof for faded old Fremont Street. It is the Sistine Ceiling of the Vatican of Mammon.

When screens of these various types, sizes, and shapes first began to colonize our everyday surroundings, they were all operated independently; your TV had no relationship to your PC, and the information that these devices displayed arrived through separate channels from very different sources. Then computers with point-and-click interfaces familiarized us with the idea that we could arrange information on a PC desktop in any way we wanted. In the not-so-distant future, as smart places become increasingly sophisticated, we will increasingly treat their display surfaces as integrated interfaces to streams of information delivered by the digital infrastructure. So you might, for example, monitor a news broadcast at postage-stamp size on a wristwatch display, then drag it to a nearby wall-sized screen when something interesting came on.

■ Out of the Box

Playwrights like Ibsen, who wanted to present realistic action unfolding as if the audience wasn't there, loved the proscenium. Its implied transparent wall created exactly the sort of relationship they wanted. But it was a much-resented impediment to playwrights and directors who wanted to immerse the audience in the action and create a greater sense of participation. This has motivated development and use of alternative theatrical setups—in particular, the open stage, the thrust stage, and the arena stage.

For similar reasons, digital media researchers have long sought ways to break out of the computer screen's rigid rectangle and to immerse us in electronically delivered information. This is not so straightforward, but it can be done. One possibility—in the near technological future—might be to employ some sort of smart wallpaper, wallboard, or paint. This would allow much more freedom in display surface configuration.

The basic idea is simple; employ some sort of material that visibly changes state under electrical stimulus, spread it across the surface, and figure out some point-by-point addressing scheme for the stimuli. As conceived by the MIT Media Laboratory's Joe Jacobson, for example, "smart paper" incorporates tiny pellets that are white on one side and black on the other, and can be flipped by an electrostatic charge.[10] Alternatively, large-scale smart surfaces could consist of addressable dots of glow-in-the-dark material. And, at coarser resolution, smart ceramic tiles or smart panes of glass might create reprogrammable mosaic patterns.

Immersive displays work very differently from framed and
bounded ones. When you concentrate on a PC screen, it normally becomes the center of your attention; everything outside
its boundaries is peripheral. But when you are fully immersed
in electronically displayed information, you can only focus on
a relatively small part of it at a time, and you are no more than
peripherally aware—perhaps barely consciously—of the rest of it.

Peripheral information is by no means unimportant; in fact,
it plays a crucial role in establishing the character of a place and
sustaining your relationship to it. When a room has a window, for
example, it provides a continuous flow of information about the
external environment—the cycles of day and night, the movements of sunlight and shadows, the succession of bright and
cloudy moments, and the alternation of dry and rainy patches.
You rarely pay explicit attention to all this, but you are peripherally aware of it, and you feel uncomfortably isolated if you are
cut off from it.

Similarly, in a crowded restaurant, you give most of your
attention to your immediate companions, but you remain
peripherally aware of the background buzz of conversation, the
blur of surrounding faces, and the constant movement of waiters.
If there is a perceptible change—a sudden hush in conversation,
for example, or a crash of plates—you may momentarily shift
your attention to the source of the disturbance. You may also
switch it when your needs change; thus you start paying more
attention to the movements of the waiters when the meal is end-
ing and you want the bill.

In a movie theater, you concentrate on the action unfolding
on the screen. But simultaneously, you are peripherally aware of
the audience reactions around you, and this is a crucial part of the
experience. And, if somebody shouts "fire," you will quickly divert
your attention to the exits.

In immersive digitally mediated environments, surfaces and
objects can be activated in subtle ways to present similar back-

ground information. Fluctuating levels of potentially important quantities such as stock prices, network traffic, pollution counts, and building energy consumption can, for example, be represented by low-level ambient noise (like rain on the roof), vibrating wires, fountains, pinwheels that spin in a "bit wind," and water ripple shadows resulting from a "rain of bits."[11] In addition, through transmission of audio and video information, peripheries can be transplanted; thus in electronically linked collaborative workspaces, you might hear the combined background noise of activity at different locations, while keeping an eye on who's there by glancing occasionally at webcam images displayed somewhere at the edges of your normal field of view.

So breaking the boundaries of the screen does far more than provide additional display area. It opens up the possibility of smart places that engage our senses and attract our attention at multiple levels.

■ Up in Lights

Smartening up the enclosing surfaces is not the only way to immerse the users of a space in electronically displayed information, and to create an information periphery as well as a center. Where the geometry of a space permits obstruction-free projection (or where shadows don't matter very much), video or laser projection provides another effective way to smear information around architectural interiors.

Projectors may be fixed in place—therefore addressing strictly limited segments of wall, floor, or ceiling surfaces—or they may be mounted on gimbals like surveillance cameras so that they can address entire architectural volumes.[12] Thus, in Pierre Wellner's Digital Desk project, for instance, a standard desk was augmented with an overhead video camera and projector, allowing paper documents to be intermingled freely with projected digital ones.[13] And in Hiroshi Ishii's metaDESK project, bottom-projection of video images onto a translucent desk surface was combined with use of small physical models and tools to control computational processes.[14]

Wall-sized video projections that are created in these ways can merge remarkably seamlessly with physical reality by presenting images at life size and extending them to the edges of the viewer's peripheral vision. Myron Krueger's Videoplace projects first convincingly demonstrated this possibility; Krueger created spaces in which full-scale video-projected "shadows" of inhabitants interacted with one another in complex and sometimes surprising fashions.[15] More recently, electronic image-merging has been used to create Hyper-Mirrors—large video walls in which life-size images of local and remote teleconference participants share the same virtual spaces.[16]

All this amounts to a radical reconceptualization of the whole idea of artificial lighting. Think of lightbulbs not as the dumb, one-pixel displays that Edison concocted, but as computer-controlled combinations of miniaturized video projectors and cameras.[17] Make them, say, 1,000 by 1,000 pixels. And think of their output not simply as photons bouncing off the walls, but as a highly structured, precisely controllable, interactive field of luminous energy.

■ Interface in Your Face: Virtual Reality

An even more daring trick is to miniaturize the video screens and position them directly in front of your eyes to yield a head-mounted stereo display.[18] In combination with head-tracking devices to keep the computed scene synchronized with your motion, and with sufficient computing power to refresh perspective views in real time, such displays can produce the convincing illusion of complete immersion in virtual 3D space. The proscenium is entirely gone; this virtual reality gear is a cumbersome, geeky-bondage fashion-felony, but it puts you right inside cyberspace.

This is, in effect, an inversion of the old Renaissance conception of the relationship of architectural space, perspective plane, and viewer's retina. For Alberti and Brunelleschi, the real three-dimensional scene created a virtual two-dimensional image—which might be traced by the artist—on the perspective plane.[19] For the user of a virtual reality system, by contrast, palpable, glowing, two-dimensional images on the perspective plane create a virtual three-dimensional scene.

There are some other technological means to the same end. You can, for example, make use of flicker glasses incorporating liquid-crystal shutters that alternately occlude one eye, then the other. Synchronized right-eye and left-eye images are then projected onto surrounding screens—typically arranged to form an enclosing cube—and the result, once again, is an illusion of actually being there in a 3D virtual place.[20]

Whatever the VR technology, though, the effect is to disconnect you from your physical surroundings and completely substitute an electronically constructed virtual environment. This has its problems, of course; it is easy to blunder accidentally into the actual walls, or fall over the chairs. From the perspective of unwired bystanders, who cannot see what you see, you just seem bong-headedly out of it. And this electronically sustained solipsism is mightily unhelpful to social interaction.

■ All Over: Augmented Reality

Fortunately, it is not necessary to mask out the physical surroundings entirely. You can, for example, incorporate prisms in the eyepieces of VR goggles; this superimposes computer graphics onto the surrounding scene, so that three-dimensional virtual objects seem to mingle with physical ones to create a new sort of hybrid architecture.[21] Alternatively, you can substitute video cameras for the prisms, and electronically mix the live video image with the synthetic computer graphics—very effective, so long as the video does not fail and leave you effectively blindfolded. The result is often called augmented reality, or sometimes—a bit more generally—mixed reality.

If the motion-tracking, registration, and superimposition techniques get to be good enough (no easy technological task, incidentally), such systems will increasingly perform the traditional architectural function of overlaying textual and graphic information onto the human habitat. Ancient buildings simply accomplished this with inscriptions and murals. The Gothic masters used stained glass. Las Vegas has favored neon lights. Product packagers plaster printed labels everywhere. Our own age has now contributed the possibility of virtual overlays created by microelectronics—perhaps a way to suck all that info-clutter off the surfaces that surround us and provide personalized information overlays whenever and wherever we need them.

With augmented reality, different inhabitants of a city might see different, specially tailored annotations superimposed on it. A tourist might view guidebook information and reconstructions of the past overlaid on historic sites—or, if in a different frame of mind, indicators of crime and accident sites. A real estate agent might look at buildings labeled with their asking prices, a construction worker might be guided by building designs correctly registered on vacant lots, and a technician might consult repair manual diagrams conveniently positioned on malfunctioning machines. A bicycle messenger might find the names of residents added virtually to doorways. And speakers of different languages might get it all in their native tongues.

■ Pixels, Pixels, Everywhere

In a world of proliferating screenspace and speakers, smart surfaces, video-projected displays, virtual reality, and augmented reality, luminous digital information is ubiquitously overlaid on tangible physical reality. As static tesserae were to the Romans, active pixels are to us. Signs and labels are becoming dynamic, text is jumping off the page into three-dimensional space, murals are being set in motion, and the immaterial is blending seamlessly with the material.

Architecture is no longer simply the play of masses in light. It now embraces the play of digital information in space.

Watch out! As the technology of smart places matures, the metaphors are biting back.

In the early days of computer graphics, we became familiar with "virtual" objects that looked like physical ones but could perform computational tasks. We learned to "paint" with virtual brushes, store digital "documents" by dragging them to on-screen "file folders," delete by means of icons shaped as trash cans, and so on. It was as if familiar physical artifacts had been sucked up off the desktop into the PC, there to live a ghostly, magically enhanced afterlife. Now, by embedding intelligence and interconnectivity in material products and creating systems of tags and sensors, we can reverse the process. We can squirt these sorts of computational capabilities back into everyday physical things; we can get the functionality without the virtuality.

In elementary form, this is already a familiar idea; supermarket products are tagged with printed barcodes, and checkout counters are equipped with barcode readers. Dragging a product past the reader produces a computational result; software that lurks behind the surface reads the product's identifying code, looks up its price in a database, and finally adds the price to the customer's total. It may also perform important ancillary tasks, such as updating a stock inventory and collecting statistical data on buying patterns.

By generalizing this principle, we can construct spatially extended smart spaces from collections of interacting smart objects. Real desktops, rooms, and other settings—rather than their electronically constructed surrogates—can begin to function as computer interfaces. We can also create some interesting physical/virtual hybrids, as in the video arcade golf simulators where you hit a real ball with a real club, then see a simulated trajectory displayed on a video screen. As a result, our actions in physical space are closely and unobtrusively coupled with our actions in cyberspace. We become true inhabitants of electronically mediated environments rather than mere users of computational devices.

■ Tags and Sensors

When we want physical objects to serve in this fashion as active elements of smart places, it turns out that we must provide them with ways to identify themselves to each other. The technology may be optical, as with barcodes and readers, with fingerprint recognition systems that open doors for authorized individuals, and with face recognition systems. It may be acoustic, as with devices that emit ultrasonic signals. And it may be electromagnetic, as with ATM cards, radio frequency identification (RFID) chips in those key chain tags that activate gasoline pumps, Sensormatic shoplifting tags, and smart tollbooths that automatically identify and later bill the owners of transponder-equipped vehicles that pass through them.

It sometimes matters not only *what* things are, but also *where* they currently happen to be, so we also need ways to track the positions of physical objects—much as screen management software tracks the position of a cursor. This can be accomplished in a variety of ways. At a large scale, where accuracy to within a few meters suffices, the Global Positioning System (GPS) satellite system, together with inexpensive, miniaturized GPS receivers, can provide coordinates of vehicles and other objects anywhere on the face of the earth; this information typically feeds into onboard vehicle navigation systems and emergency service call systems.[1] At urban and architectural scales, grids of terrestrial transceivers can keep track of vehicles and cell phones. Within buildings, various pressure-sensitive and motion-sensitive, electromagnetic, optical, and acoustic sensors can follow the movements of people and artifacts—allowing calls and messages to be forwarded automatically, for example.[2] And for pinpoint accuracy at a small scale, the electromagnetic and ultrasonic techniques used in 3D digitizers can serve effectively.

Some smart objects require specialized sensing capabilities, as appropriate to their particular roles. They may be equipped, if necessary, with cameras and microphones as "eyes" and "ears." They may incorporate temperature and humidity sensors. They might watch out for tiny traces of explosives, drugs, or pollutants. There might be miniature accelerometers to detect motion, piezoelectric detectors for forces and stresses in structural elements, micropower impulse radar (MIR) to measure distances and fuel levels, electric field sensors for capturing gestural information,[3] and digital compasses to track orientation. They might even make use of living cells as detectors for hormones and microorganisms. The list is potentially endless.

Like living organisms, smart objects will sometimes need to figure out what is going on around or within them by integrating sensory inputs from multiple sources.[4] In order to respond to a child, for example, a smart stuffed toy might sense both sound and motion. To monitor, interpret, and respond to an occupant's commands, a smart room might collect audio information from several microphones, video streams from multiple cameras, and occupant locations from a smart carpet or other position-sensing systems. This allows cross-checking of information, and eliminates many potential ambiguities.

For truly ubiquitous use, the tags and sensors embedded in manufactured products need to be tiny, robust, very inexpensive, and low-powered; as Neil Gershenfeld has observed, we need to be able to compute anywhere for pennies.[5] Here the technologists are beginning to deliver, though there is still a long way to go.[6] Video cameras, for example, are evolving into single-chip devices costing only a few dollars; they can become cheap "eyes" for almost anything. Micro-electromechanical systems (MEMS) technologies allow very small-scale sensor fabrication. And MEMS devices can become so minute that they can be powered by vibrations or solar energy, dispensing with external power supplies and batteries.

In general, new tag and sensor technologies allow objects to become aware of one another and to begin to interact. This is the first, elementary step toward artificial ecosystems and societies of smart stuff.

■ **Embedded Intelligence**

In order to process information and respond, smart artifacts need not only sensors, but also embedded memory and machine intelligence.

Although you may not have noticed if you aren't particularly alert to it, unobtrusive, onboard computers are now commonplace in vehicles, appliances, and even toys. Your automobile has sophisticated digital systems to control braking and other functions; indeed, these probably account for more of the cost than the engine and the power train combined, and they consume so much electricity that they will probably force a move from 12-volt to 42-volt batteries. Your microwave, your dishwasher, and your clothes washer incorporate more processing power than advanced computers of a few decades earlier. Television receivers and cell phones are packed with digital circuitry. Complex film-loaded cameras are giving way to digital electronic ones with almost no moving parts. Programmable card key systems are replacing mechanical locks and keys on doors. Microsoft Barney, the annoying stuffed toy spun off from the annoying kids' TV character, has a speech chip and a motion controller implanted under his purple polyester pelt. Dissect a Furby and you get an electronics lesson.

This extends a revolution in product design that has quietly been cooking since the appearance of the first microchips in the 1960s; mechanical and electromechanical subsystems account for a steadily decreasing proportion of such a product's functionality and cost, while digital electronic subsystems take up a correspondingly increasing share. By the mid-1990s, as a result, microprocessors embedded in specialized smart artifacts outnumbered personal computers by an astonishing factor of one thousand.[7]

As chips become smaller, cheaper, more capable, and more robust, and as their electric power requirements decrease, this wholesale invasion of manufactured products by digital intelligence will continue. There will be local processing power and memory available wherever it may be needed, for any purpose whatsoever. Eventually, we will cease to conceive of computers as separate devices, and begin to regard machine intelligence as a property that might be associated with just about anything.

We will increasingly inhabit a world of things that don't just *sit* there, but actually *consider* what they are supposed to be doing and choose their actions accordingly.

■ Instant Networking

How can we actually configure such smart components to transform our immediate surroundings into smart spaces?

In the PC era, the answer seemed simple. You assembled computer resources in a room by plugging miscellaneous peripheral devices into a CPU box, then loading shrinkwrapped software. But this process became increasingly cumbersome as intelligent objects diversified and proliferated. All those tangled wires and blinking boxes were just too much trouble. They had to go!

An obvious first step was to replace jacks and cables with a universal short-range radio linkup among neighboring electronic devices. This could be accomplished by equipping them all with miniaturized, high-frequency, low-powered transmitters and receivers. The Bluetooth technology specification, introduced in the late 1990s by a consortium of major electronics firms, opened up this possibility by providing a workable and widely supported standard.[8] When Bluetooth devices come in close proximity, they automatically detect one another and establish a network connection.

Unfortunately, though, physical interconnection of devices does not suffice to make them *work* together. (You probably know this well if you have ever tried to hook a new printer to your PC, or connect your laptop to a conference room video projector.) You also need some simple, automatic, foolproof way to handle the hardware compatibility issues that inevitably arise. The devices have to communicate through some common digital language. Providing this lingua franca is the function of "network dial tone" software, such as Sun Microsystems's Jini; it is designed to make all the resources of a network immediately available to any new device that links in, while simultaneously allowing that device to function as a new network resource.[9]

With wireless connection and automatic assurance of hardware compatibility, electronic devices can click together as effortlessly as Lego blocks. Networks become less like fixed plumbing and more like ad hoc furniture arrangements configured for particular, temporary purposes.

■ Rhizomic Software

Once a piece of smart hardware is part of a network, it can potentially download any software or connect to any network services that it may need. Thus we might imagine the capabilities of smart places being configured on the fly, as required for particular purposes— a radically new process of wide-ranging, search-engine-enabled, electronic bricollage.

In practice, some arcane but crucial issues of software style and structure need to be sorted out before this attractive idea becomes workable. In particular, it helps a great deal if code is organized not in the form of huge, monolithic systems but as collections of modular, reusable, recombinable components containing both executable statements and data; this is the underlying principle of object-oriented programming, and of languages such as C++.

Furthermore, these code components are most widely useful when they can run not only within the operating system and machine environment for which they were written, but within any computational setting. The Java software environment, for example, makes this possible by providing "virtual machines" that run on top of particular hardware and operating systems to provide uniform execution environments.[10] All this is very inefficient, but that matters little in an era of inexpensive, powerful processors and ample memory.

More radically, code to perform particular tasks may be encapsulated in the form of autonomous software *agents*.[11] Like itinerant performers, these may roam a network in search of sites to do their thing.

By the end of the 1990s, it was becoming clear to industry insiders that smart devices, ad hoc networking, and modular, relocatable software were combining to create much more flexible computational environments than those of the past. The 1960s and 1970s had been the era of centralized timesharing systems, the 1980s and early 1990s had seen client/server systems, the Internet, and the World Wide Web, but the new century would be characterized by interconnected smart stuff everywhere. University and industrial research laboratories began to flesh out the details; MIT's Media Laboratory initiated an ambitious Things That Think project, MIT's Laboratory for Computer Science worked on a prototype technology called Oxygen, Hewlett-Packard announced its commitment to "service-centric computing," and Sun pushed Java and Jini.

■ **Form Fetches Function**

When software becomes footloose in this way, and services are there for the connecting, we can no longer expect the functions of things to be as stable and predictable as they once were. Now, a display screen on a wall might serve successively, according to our whims of the moment, as a clock, a television, a stock ticker, a portrait of a loved one, and a remote nannycam monitor. A single handheld device might play the roles of cell phone, pager, personal digital assistant, and television remote control. A simple plastic rectangle might function as a credit card, a wallet filled with digital cash, and a door key. An ATM machine (unlike an old-fashioned branch bank) might offer the services of many different banks and other financial institutions, depending upon the identities and needs of particular customers.

Nor can we expect these functions to be localized. Any smart, networked device becomes a tangible, local delivery point for an indefinitely extensible, globally distributed pool of resources and services. Some of these may be embodied as pieces of hardware somewhere, some may be accomplished by execution of software, some may be performed by actual people, and you mostly do not know or care which is which. And, if network connections are sufficiently fast, it scarcely matters whether a task is performed locally or on a processor that happens to be available on the other side of the world.

Consequently, product designers and architects face some new sorts of design quandaries. Should they build multipurpose hardware, such as multimedia personal computers, or should they create families of interacting, single-purpose devices like cell phones, digital cameras, and portable electronic books—information appliances that fragment and disperse the functions?[12] Which capabilities of a system should be built into hardware and which should be provided by software? Which software functions should be permanently resident in a device and which should derive, as necessary, from interconnections and downloads? In the end, some affordances will derive from material structures and mechanisms, some from resident code, some from software and services sucked down a wire as needed, and some from interactions of all of these.

In the design of smart things and places, form may still follow function—but only up to a point. For the rest, function follows code. And if you need to alter these code-enabled functions, you don't rebuild, reshape, or replace material components; you just connect, fetch, and load.

■   Consult the Genius of the Place

Curiously enough, there is venerable precedent for these notions of embedded intelligence and sensorily aware, responsive objects and spaces. The ancient Romans believed that each particular place had its characteristic spirit—its *genius loci*—that might manifest itself, if you watched carefully for it, as a snake. They had the right idea, but not the necessary technology.

For us, equipping a place with its genius has simply become a software implementation task. Lines of code can supply every electronically augmented environment with a tailor-made, digital genius that makes its presence felt through input devices and sensors, displays, and robotic actuators. It can respond to the needs of its inhabitants, adapt to changes in its surroundings, and—by making use of its network connectivity—focus global resources on current local tasks. By virtue of the rules that it encodes, it can facilitate some activities and discourage or exclude others. It can even enforce ethical and legal norms.

Code is character. Code is the law.

What will smart places do for us?

They will, of course, collect and spit out information—much as computers and telecommunication devices have always done. More importantly, though, they will attend, anticipate, and respond to our daily needs in a vast variety of new ways. And they will become delivery points for a still-unimaginable range of services made available by providers scattered around the globe.

■   Wear Ware

Most intimately, there will be close-fitting networks of implanted, wearable, and pocket devices to attend to our most immediate, ongoing requirements for maintenance of bodily health and comfort, for self-representation and identification, and for remote communication.[1]

Our clothing and accessories will be dense with bits. There may be more lines of code in your shoe that on the disk of your current desktop PC. If this seems a bit too Buck Rogers to credit, try the experiment of emptying out your pockets, handbags, and briefcases, counting up all the objects that record, store, display, or process information in some way, and imagining their replacement by smaller, lighter, very much smarter digital equivalents. This replacement process began with watches and cell phones, and it will continue.

There is plenty of room for the necessary intelligence in footwear, belts, jackets, hats, wallets, handbags, briefcases, wrist straps, and buttons. Gloves and other close-fitting clothing can serve as gesture sensors. Tiny, lightweight CCD arrays and microphones can augment your eyes and ears. Miniature displays can be kept in pockets, strapped to wrists, and built into eyeglasses. Information can be discreetly whispered into your ear as you need it, or superimposed on a scene by means of smart eyeglasses.

You will be able to take lots of data onboard. Simple credit cards and ID cards can grow up into vastly more sophisticated smart cards with large digital memories and onboard processing power. Coins and banknotes can be replaced by digital cash—encrypted bundles of bits securely stored somewhere on your person. Identifying and authorizing tokens—badges, business cards, driver's licenses, tickets, passports, visas, and door keys—can migrate from paper and metal to miniature digital cards, controllers, and transceiver badges.

And there will be myriad gadgets to deliver whatever specialized personal services your condition and lifestyle may require. Your health may demand prosthetic devices such as hearing aids and pacemakers, medical monitoring devices, and programmable and remotely controlled medication dispensers. If you ride a motorcycle or ski, you may need intelligent, dynamic protective devices such as inflatable neck braces. If you are a diver, pilot, firefighter, or handler of toxic materials, you may depend upon protective suits and specialized life support devices. Your more mundane daily activities may require cellular telephones, pagers, personal digital assistants, and audio and video entertainment devices. Even your jewelry may be programmed.

■ Body Nets

Many of these handhelds and wearables, such as smart cards, digital wallets, and digital address books, will not require continuous network linkage; these will depend on their internal memories, and will operate in plug-in-and-load-up mode. Others, like watchband pagers, will need bursts of connectivity. Yet others, like personal radios and televisions, will receive or transmit constantly.

These electronic organs may communicate among themselves, as necessary, by means of circuitry woven unobtrusively into clothing, and connections made by buttons and snaps. They may transmit digital data (quite harmlessly) through your very flesh.[2] They may even intercommunicate indirectly via microcellular transceivers in the surrounding architectural environment.

In any case, their intercommunication capabilities will allow them to act together as a versatile and efficient system that serves a wide variety of purposes. Thus, for example, grasping a smart card with your fingers might cause your wrist device to display the amount of digital cash the card currently contains. A signal from a medical monitoring device at one location on your body might trigger release of medication at another. And you might transfer files from one bodynet to another simply by holding hands.

By the late 1990s, laboratory experimentation with wearables and bodynets had intersected with cultural theorizing of the extended and transformed body (as exemplified by the influential work of Donna Haraway),[3] with productions of body artists like Stelarc, and with the checkbooks of venture capitalists. Consumer electronics giants were experimenting with wearable digital products; Seiko, for example, put out a wristwatch wireless messager. The hopeful start-up companies were appearing.[4] MIT hackers, decked out in their digital cyborg gear, were appearing in the fashion pages of the *New York Times*. And Gordon Bell was predicting: "By 2047, one can imagine a body-networked, on-board assistant—a guardian angel that can capture and retrieve everything we hear, read, and see. It could have as much processing power as its master, that is 1,000 million-million operations per second (one petaops), and a memory of 10 terabytes."[5]

You will be certain that this cyborgian stage of the digital revolution has truly arrived when breadbox-sized computer boxes mostly fade from view, and you just put on your digital devices and network connections like boxer shorts.

■ Appliance Intelligence

At the next scale up from wearables—that of furniture, permanent equipment, and desktop devices—your immediate surroundings will unobtrusively be imbued with electronic intelligence.

You will engage smarter and smarter boxes, vehicles, appliances, and toys to perform specialized tasks in particular contexts—ATM machines in public places for banking, point-of-sale computers for processing retail transactions in stores and supermarkets, electronic information kiosks in transportation terminals and building lobbies, desktop devices and printers for information work in studies and offices, videoconferencing systems in meeting rooms, onboard navigation systems in vehicles, speech recognition and synthesis systems in the nursery, programmable control systems in kitchen and laundry appliances, and many more yet unimaginable.

It is useful to put this development in a very broad historical perspective. Urbanization allowed us to accumulate nonportable possessions—to populate our habitats with furniture, pictures, rugs, lamps, pianos, tableware, and all the sorts of stuff you load into a moving van when you change houses. Then mechanization took command. The industrial revolution inserted machinery into many familiar artifacts, created new mechanized products that nobody had ever dreamed of before, and produced a world that required the attention of mechanics and service technicians. Electrical networking and the proliferation of small electric motors furthered this process, ushered in the era of plug-in electric appliances, and provided our daily existence with an electromechanical middleground. Now, digital networking and small electronic processors are transforming not-so-smart appliances into very much cleverer robots.

We have gone from the writing desk to the mechanical typewriter, the electric typewriter, and finally the word processor. Cash drawers became cash registers, then point-of-sale computers. The sketchbook morphed into the film-loaded camera, then the digital camera. The craftsman's tools gave way to steam-powered and electric-powered factory machines, then to industrial robots. And the horseless carriage was the first step toward the pilotless aircraft.

■ Electronic Teamwork

Unlike earlier generations of computer-controlled appliances, though, future generations will rely on their communications capabilities and network connections. They will be members of electronic teams. As in sports teams, the individual devices will have specialized roles and positions.

They will interact with smaller-scale bodynet devices, with their counterparts elsewhere in the immediate surroundings, and with larger-scale systems. This means that their affordances are not limited by the direct capabilities of their onboard hardware and software. They may suck in necessary information, as required, from distant sources. They may pump information out to remote devices that provide numerous additional functions. And they may temporarily appropriate remote memory and processing power to assist with particularly demanding tasks.

Once, for example, you snapped a baby picture, took the film to a processor, and mailed the picture to your mother. Now, you can capture the image with a digital camera, point the camera at a PC to transfer the pixels wirelessly to disk storage, immediately distribute to your entire extended family through the Internet, and leave it to them to make prints, if they like, on their home printers. You transfer bits rather than atoms, and you perform the various necessary functions at different sites from those established in the days of mechanical shutters, silver-based emulsions, and darkrooms.

Similarly, you once put a metal coin in a mechanical meter when you parked your automobile. You had to have a pocketful of change. Today, in some places, you can pay by swiping a smart card through an electronic meter. In future, the meter will probably communicate wirelessly with a transponder in your car, and automatically generate a charge for which you are billed at the end of the month. You will not need to think about it at all.

Not so long ago, when your car broke down, you had to trudge to the nearest public telephone to call the tow truck. These days, you are more likely to call from your digital cell phone. And increasingly, automobiles are being equipped with advanced computer and telecommunications systems that locate them using GPS tracking systems, automatically diagnose problems and summon service, consult computerized service records, and even allow some adjustments and repairs to be carried out remotely.

If you walked into an early-electronic-era conference room to
make a presentation, you had to plug your laptop into the video projector, boot up the system, and hope that everything turned out to be
compatible. Before too long, your laptop will spontaneously join the
local network wherever you happen to be, and devices like video
projectors and printers will simply announce themselves on the desktop and offer their services. So will the light switches, the controls for
the window shades, the air conditioner temperature dial, and the
video remote.

Thus the old dream of a robot-serviced future is finally material-
izing—but in the form of geographically distributed assemblages of
diverse, highly specialized, intercommunicating intelligent artifacts,
not those cadres of clanking multipurpose humanoids imagined in
the late industrial era by Karel Čapek and Fritz Lang. This has evoked
the usual responses to such advances in the functional range of appliances and gadgets—breathless, won't-it-be-cool scenarios of never-
lift-a-finger future ease, countered by equally predictable rejoinders
that it's all just a back-to-mama fantasy projected onto the latest crop
of new machines.

But as in the past, both of these intellectual reflexes kick in
wildly wrong directions; abundant machine intelligence is just like
steel, plastic, and the electric motor—a useful addition to the designer's repertoire, to be used as appropriate, in conjunction with other
materials and components, to create varied physical products that
serve our needs and satisfy our desires. The cleverest, most successful
designs will not parade their computational capabilities. Your cell
phone, for example, internally performs some astoundingly complex
operations, and carries hundreds of thousands of lines of code to
provide it with the capacity to do so. Furthermore, it interacts continuously with a sophisticated surrounding environment of cellular
transceivers. But all this is transparent to you. You just see that it performs a straightforward function reliably and effectively.

COMPUTERS FOR LIVING IN

■ **Buildings with Nervous Systems**

These developments suggest a new evolutionary stage for architecture. Our buildings will become less like protozoa and more like us. We will continually interact with them, and increasingly think of them as robots for living in.

In the distant past, they were little more than skeleton and skin. Following the industrial revolution, they acquired elaborate mechanical physiologies—heating-ventilation-air conditioning (HVAC) systems, water supply and waste removal, electrical power and other energy systems, mechanical circulation systems, and a wide variety of safety and security systems; pretty soon, these evolved to the point where they were responsible for the bulk of a building's construction and operating costs. Today, in the wake of the digital revolution, they are getting artificial nervous systems, sensors, displays, and computer-controlled appliances; the structure becomes a chassis for the sophisti-cated electronic systems that play a rapidly growing role in responding to the requirements of the inhabitants.

Integration of the necessary digital telecommunications wiring raises much the same design issues as provision of electrical wiring and older forms of telephone wiring. You need vertical and horizon-tal distribution through some combination of walls, floors, ceilings, furniture, and special chases and trays, together with accessible junc-tion boxes and wiring closets. And you need a system of modular jacks to provide convenient plug-in access wherever needed. But these issues loom larger since the overall amount of wiring increases dramatically, and since the pace of technological change requires flex-ibility and easy access everywhere.

Wireless transceivers in ceilings and elsewhere can eliminate the cables running from plugs to appliances, but they do not remove the need for a well-designed, flexible wire management system. Even if they have wireless data connections, computers and other digital devices still need electric power supply. And, since electromagnetic spectrum is a scarce resource but wired capacity can be expanded indefinitely, cables are likely to remain the most efficient means of supplying high-speed connections in densely inhabited spaces.

In the end, though, the precise character of a building's digital plumbing is a relatively unimportant technicality. What really matters is its pervasiveness—its capacity to collect and deliver bits *anywhere*.

■ Livables

Just as light fixtures, HVAC diffusers, and other such elements have found their natural places in architectural settings, so will the new electronic organs that a building's artificial nervous system interconnects—its sensors, displays, projection surfaces, and robotic actuators. As this evolution unfolds, the distinction between building and computer interface will effectively disappear. Inhabitation and computer interaction will be simultaneous and inseparable.

Mark Weiser's Ubiquitous Computing project, at the Xerox Palo Alto Research Center in the early 1990s, provided one of the first convincing glimpses of this possibility.[6] Within the interior space that Weiser created, office workers wore wireless transponder pins that allowed a computer to track their locations. The environment was populated with wearable, handheld, and part-of-the-furniture display and interaction devices. These devices were all interconnected to create a single, dispersed, interactive interface. The inhabitants became, in effect, living cursors; information that they needed automatically followed them from place to place, and showed up on whatever display device happened to be convenient at the moment. And the building always knew, from moment to moment, exactly where to forward their phone calls and their email.

Around the same time, George Fitzmaurice's spatially aware palmtop computers vividly demonstrated the potential interrelationship of wearables and livables.[7] These handheld devices had location and orientation sensors, and they delivered information that was relevant to current locations or nearby objects. Thus they turned whole environments into spatially organized information fields. Point your palmtop at a malfunctioning appliance, for instance, and it might identify and summon an appropriate service technician. Point it at a product in a showroom and get technical specifications. Or point it at a museum exhibit and get catalog copy.

As designers explore these new possibilities, they will find themselves questioning old assumptions about what goes where—in particular, the traditional assignments of functions to handhelds and wearables, to the permanent elements of local settings, and to remote sites. Do you store your personal records on your body, on a computer in your house, or on a remote server? Do you keep books and records on hand, in your living room, or do you download digital versions whenever you need them? Do you sketch on a handheld surface, or on a wall-mounted electronic whiteboard? Do you control the lights and appliances in your house through traditional switches and dials mounted on the walls, through "soft" control panels on conveniently located video screens—much like the "glass" cockpits that have supplanted complex arrays of instruments in modern aircraft—or through handheld wireless devices like TV remotes?

■ Intelligent Resource Consumption

Smart buildings will not only be highly responsive to the needs of their inhabitants, but they will also be very intelligent consumers of resources in doing so.[8] They will be programmed to adapt themselves not just to variations in internal demands and external climatic conditions, but also to ongoing changes in the prices of the various utilities that serve them. This will allow utility companies and other suppliers to more effectively manage demand by adopting dynamic pricing strategies.

Consider electricity supply. The earliest idea was to have a large central generating plant that had a monopoly on pumping electric power out to consumers in its supply area. Simple meters sufficed to measure consumption; you got your monthly bill after the meter reader came. Then there were electricity grids, with multiple plants supplying power in different amounts, at different times, and at different prices; electric utility companies got into the business of buying power and distributing and reselling it to consumers. Now the trend is to create highly decentralized grids with large numbers of relatively small suppliers—possibly including buildings that generate excess solar or wind power from time to time, and output it to the grid.[9] Furthermore, utility companies have long since discovered that demand fluctuates a great deal, and that it is in their interest to try to manage this by varying prices—for example, by introducing peak and off-peak rates. The market isn't so simple any more.

Under these conditions—for efficiency and equity—prices should be updated as dynamically as possible. Smart buildings should then be programmed to respond appropriately by adjusting their demands—consuming as little power as possible when prices are high, and performing intensively power-consuming tasks when prices are lowest. This is feasible because they usually have functions—such as running a home dishwasher or cooling an empty office building after a warm day—that can take place at different times and rates without causing inconvenience. So they can shop for the best times and prices. They might also be connected to multiple supply grids, and have the ability to switch among them according to current costs.

In general, smart devices and intelligent environments will be programmed to forage intelligently for the supplies and conditions that they need in order to operate. This allows creation of more sophisticated markets, leading to more efficient use of scarce resources. Gardens can be watered automatically when other demands on the water supply system are lowest. Smart cars can take account of dynamic road pricing when choosing routes.[10] Computer systems can download large files from the Internet at off-peak times. If the technical and social kinks can be worked out of the idea of meta-computing—automatically borrowing idle processors on a network to share the load of large computational tasks—we can even begin to think of the Internet as a vast computer-power grid with dynamic pricing of machine cycles.[11] Electronic intelligence, embedded everywhere, creates the necessary interfaces between producers and consumers, and allows us to rethink the ways that even the most mundane utilities are organized and run.

Fittingly, digital information itself may be the commodity most suited to dynamic utility pricing and smart foraging. The value of information often decays with time; yesterday's newspaper is not worth as much as today's, stock price information is useless if it is not timely enough, emergency medical information is ineffective if it does not arrive when needed, and any scarcity value possessed by an item of information quickly disappears as it is duplicated and distributed through a network. So dynamic pricing of network-distributed digital information, based on its timeliness and relevance in specific contexts, provides one possible solution to problems created by collapse of the "intellectual property" approach to controlling and marketing information. The idea is to charge high prices for really hot stuff, and let everything else be inexpensive or even free.

■ Adaptive Behavior

Whether the operation of such smart resource-consuming systems is fully automatic, or whether it relies on displays of salient information and human attention, may turn out to be largely a matter of taste—like the choice between manual and automatic transmissions in automobiles. It may depend upon whether you enjoy driving or just want to devote your attention to something else.

One thing seems certain, though; few of us really want to program even the simplest of devices, like VCRs, microwaves, telephone answering machines, and cameras—much less our homes, offices, or classrooms. No doubt the notoriously terrible interfaces and incomprehensible instruction manuals of these devices are partially to blame for our distaste. But it is more fundamental than that. We should not have to explicitly instruct our appliances and environments at all; if they are *really* so smart, they should be able to learn what we require of them by watching us. Like the best of waiters or personal secretaries, they should be able to anticipate our needs before we are even consciously aware of them. Otherwise, these complicated gadgets are often more trouble than they are worth.

How smart, then, does a washing machine need to be? Perhaps it should automatically analyze clothing stains, mix chemicals, set the rinse and agitation cycle, and reorder supplies over the Internet. Maybe it should learn when you like to have fresh laundry ready, note the prevailing patterns of electrical power pricing, and time its operations accordingly.

What about wallboard? Maybe smart sheetrock should observe your comings and goings, automatically create predictive models of your behavior, and provide your house with the capacity to perform its environmental control chores accordingly. It might even learn to distinguish between the differing environmental needs of your teenage daughter and your aged grandmother, and to take account of who was actually at home.[12] If it got to be good enough at these games, it could satisfy your HVAC and lighting needs while cleverly minimizing energy costs. The longer you lived there, the better it would get to know you, and the better it would do.

All this becomes feasible if effective machine-learning mechanisms can be built into smart devices and places. One of the most convincing demonstrations of the possibilities, so far, is Michael Mozer's "adaptive house" in Boulder, Colorado.[13] Mozer's house (actually a retrofitted former schoolhouse) incorporates an elaborate array of sensors that monitor internal temperatures, ambient light levels, room-by-room sound and motion, the openings and closings of doors and windows, outdoor weather conditions, boiler temperature, and hot water usage. Its heating, ventilation, and lighting systems are computer-controlled. A neural network system tracks occupant movements and behavior, predicts comings and goings and room occupancies, and infers rules of operation that appropriately balance occupant comfort and energy conservation.

■ Reconceiving Construction

As buildings evolve in the directions represented by these new ideas and pioneering experiments, construction materials, products, and processes will change. Steel and concrete will still be important, but they will be joined by silicon and software.

The buildings of the near future will function more and more like large computers with multiple processors, distributed memory, numerous devices to control, and network connections to take care of. They will continuously suck in information from their interiors and surroundings, and they will construct and maintain complex, dynamic information overlays delivered through miniature devices worn or held by inhabitants, screens and speakers in the walls and ceilings, and projections onto enclosing surfaces. The software to manage all this will be a crucial design concern. The operating system for your house will become as essential as the roof, and certainly far more important than the operating system for your desktop PC.

Consequently, a growing proportion of a building's construction cost will go into high-value, factory-made, electronics-loaded, software-programmed components and subsystems; a correspondingly decreasing proportion will go into on-site construction of the structure and cladding. There will be fewer discrete components, fewer complicated mechanisms, fewer moving parts to wear out and break, and much more reliance on software and solid-state circuitry to provide the necessary functions. For convenient repair, replacement, and upgrades, these sophisticated new components will need to be modular and removable; they will snap into place like the boards inside a PC, or simply plug in where required. As they become denser with wiring and electronic devices, they will become more like large-scale printed circuit boards than dumb wallboard.

Miniaturization will allow us to exploit redundancy. Instead of relying on one light fixture to illuminate a room, we might have thousands of independent pixels; it does not matter if a few burn out. And instead of providing one large fan for ventilation, we might substitute wall panels with hundreds of thumbnail-sized turbines.

These hardware and software components will become obsolete at very different rates, and repair, maintenance, and renovation strategies will have to provide for this. Simple, robust, long-life components will form a permanent chassis. Replaceable electronic devices will plug into this. Software will continually, automatically upgrade itself via

network connections. And maintenance providers will make extensive use of remote monitoring to detect problems, analyze them, and
invoke service procedures as needed.

All this will mean that some new trades will appear on construction sites. Networking specialists, hardware technicians, and software
hackers will increasingly join the steelworkers, concrete guys, carpen-
ters, plasterers, painters, plumbers, tin-bashers, and electricians.

■ The Knee Bone Connected to the I-bahn

Smart places, at the various scales we have now considered, nest one
within the other like Chinese boxes. They form approximate hierarchies, with constant exchanges of information across the interfaces
between the levels.

Think of your brain, in the near future, as a kernel surrounded
by successive electronic shells. The innermost is your bodynet, which
employs sensors and controls that detect small-scale gestures and subtle bodily states, together with displays, speakers, and tactile devices
placed in close proximity to sensory organs, to transfer information
back and forth across the carbon/silicon divide.

Your bodynet frequently finds itself situated within smart houses,
hotel rooms, offices, stores, automobiles, airplane cabins, and other
wired settings. Such settings are rich in connection points for your
bodynet devices—either wireless transmitters and receivers or jacks
for cables—and they are populated by information appliances that
collect and process local information, while simultaneously importing
information from the global networks. (Television receivers controlled by handheld remotes, and cordless telephones, are the humble
forerunners of these information appliance systems.) Displays may be
larger, speakers may be louder, and viewers and listeners may be
groups as well as individuals.

Next in the hierarchy are the electronic territories of social
groups such as families, companies, university communities, and professional associations. Sometimes they correspond to physical territo-

ries, as in the case of local-area networks in corporate facilities and on university campuses, but they may also be diffused over wide areas. Access to them may be controlled physically or by means of passwords, firewalls, and filters.

Finally, there are the large-scale territories of terrestrial cellular systems, the footprints of geosynchronous communications satellites, and the LEO global satellite systems. These engulf vast tracts of land and sea, and are rapidly transforming the entire surface of Spaceship Earth into one all-engulfing smart place—a global market, distribution system, and agora.

■ Smart Cities of the Twenty-first Century

This proliferation of nested smart places will eventually produce a new type of urban tissue, and in the end it will radically reshape our cities.

To an excellent first approximation, the places that a city contains, the activities that those places support, and the tissues that result derive their characters from the affordances of the networks that serve them.[14] By putting in sophisticated water supply and sewer networks, for example, ancient Roman engineers succeeded in creating densely packed systems of (relatively) *sanitary* places. When the industrial revolution brought gas and electric networks, cities everywhere became collections of *illuminated* places and could extend their activities around the clock—liberating themselves from the ancient bondage of the diurnal cycle. Furnaces, pipes for hot water and steam, and ducts for air enabled creation of centrally *warmed* places, and made urban life far more comfortable in cold climates. By contrast, air conditioners plugged into the power grid allowed cities like Phoenix to develop as far-flung constructions of *cooled* places— among which people shuttle in their chilled vehicles. And Alexander Graham Bell opened the way to a world of *connected* places.

Civilization has its discontents, and each of these transformations has had its downsides. Furthermore, the short-term effects have usually been to increase gaps between the privileged and the not-so; you can be sure that the rich and powerful were always the first to get piped water supply and sanitation, electric light, efficient heating and air conditioning, and telephones.[15] But the longer-term effects of these environmental improvements have been life-enhancing, and few of us—even the most hardened technoskeptics—would want to turn the clock back.

Digital networks continue this story. We will characterize cities of the twenty-first century as systems of interlinked, interacting, silicon- and software-saturated *smart, attentive,* and *responsive* places. We will encounter them at the scales of clothing, rooms, buildings, campuses and neighborhoods, metropolitan regions, and global infrastructures.

The newly dense and abundant interlinkage provided by growing numbers of smart places embedded in the expanding digital telecommunications infrastructure is already changing the spatial distribution of economic and social activities—and hence the life and forms of our cities—by enabling dispersed, decentralized transactions among people and organizations, and by facilitating new, flexible, and efficient systems of production, storage, and distribution.

It is creating vast virtual marketplaces for labor, services, and goods that provide sellers with access to more potential buyers, but at the same time give buyers more choices and more detailed, accurate, and up-to-date price and availability information. By reshaping distribution systems, it is also shifting sites of consumption. And, by supporting ongoing interaction through telecommunication, it is producing and sustaining far-flung communities of practice, common interest, and common language and culture.

■ **The Displacement of Place?**

These new arrangements present us with new choices—frequently very attractive ones—and clearly create formidable competition for familiar, place-based enterprises and institutions. Do we continue to commute to the office or begin to telecommute from home? Do we support our local bookstores or order from online catalogs? Do we download videos for private viewing or go out to the theater? Do we give our attention and loyalty to our distant, electronically connected friends and colleagues or to our immediate neighbors—with whom we may actually have less in common?

But long-established settlement patterns and social arrangements are remarkably resistant to even the powerful pressures for change; mostly they transform slowly, messily, unevenly, and incompletely, and human nature hardly alters at all. So the outcome of this emerging competition will not simply be some eye-popping, out-of-the-blue, all-encompassing Tomorrowland; there will be lots of local specializations, contradictions, slippages, and singularities within the reconfigured world system. Global forces will contend edgily with localized resistances. New locational freedoms will be balanced against sunk investments in particular localities. Differences in topography, climate, and regional resources will certainly still matter. Unprecedented technological opportunities will be constrained by long-standing legacies of history. Technological development will interact with social and political interests, economic strategies, and cultural values in complex and sometimes surprising ways to produce a rich diversity of places and neighborhoods.

■ Reconfigured Homes

In particular, the loosening of locational imperatives by means of electronic interconnection will not peg the needle at the logical extreme. It will not turn us all into rootless, laptop-toting, cell-phoning nomads. Far from it.

Most of us will still want more or less permanent places of our own, and will choose to live in small groups of those whose company we particularly cherish—in twosomes, ménages-à-n, nuclear families, extended families, and all manner of postnuclear reshuffles and extranuclear inventions. Home, in a variety of new configurations, will be where many hearts remain—and it will be where a growing number of other things end up as well. It will become a renewed focus of architectural attention and innovation as it integrates new functions and services.

Whereas the industrial revolution forced the separation of home and workplace, the digital revolution is bringing them back together; we will see an increasing amount of electronically enabled home work, and correspondingly burgeoning demand for space in the home to accommodate it.[1] And for those who want to spend more time at home with their loved ones (or are compelled by age or infirmity to do so) electronic delivery of services—from online grocery shopping to electronic medical monitoring—will provide the necessary means.

This does not mean that the majority of us will become full-time, stay-at-home telecommuters, and that traditional workplaces—

particularly downtown offices—will simply disappear.[2] Despite decades of interest in the possibility of telecommuting, there is little evidence that it will take over to such an extent.[3] But we will certainly see increasingly flexible work schedules and spatial patterns, and many people will divide their time, in varying proportions, among traditional types of workplaces, ad hoc work settings that serve while they are on the road, and electronically equipped home workplaces.

All this is consistent with the basic human need to *belong* somewhere in particular. There is no reason to believe that this need will disappear as a result of increased electronic interconnectivity, or that all places on earth will suddenly begin to seem the same. We will not have a world where there's no there anywhere. Just the opposite, in fact. We will increasingly take advantage of digital telecommunications technology to stay more closely in touch with places that are particularly meaningful to us when we travel.

There will still be some place we call "home." And when we are far from it, we will continue to call home.

■ Rethinking Planning and Zoning

These wired homes of the twenty-first century will require more than additional space to accommodate their wider ranges of functions. The internal subdivision and organization of space will also need rethinking.[4]

There is, in particular, a potential conflict between the ideas of home as an activity center and as a refuge, and resolving this will require careful planning. So will reconciliation of privacy needs with the presence of networked microphones and video cameras. Ad hoc solutions like converting spare bedrooms to computer-equipped studies may suffice for a while, but not in the long run.

We will eventually have to invent new housing types—in many ways modern equivalents of the barbershop in Little Italy, behind which the barber's family lived. For workable prototypes, we might look to the *machiya*[5] in Kyoto's artisan districts, or to the old Peranakan shop-houses of Singapore, in which merchant families dwelt above their stores and the distinction between workspace and zones of retreat to family life was elegantly maintained by the separation of levels. In American and European cities, artists' lofts have demonstrated the potential advantages of living and working in the same place, and so provide another useful model.

In pursuing these strategies, we will find that we can exploit economies of scale in new ways. Just as large, traditional apartment buildings have frequently been able to support health clubs and doormen, so live/work complexes will be able to provide receptionists, conference rooms, and specialized equipment that otherwise would not be available in home offices.

We will also need to reexamine traditional approaches to land use zoning, which presume that workplaces generate noise, traffic, and pollution and hence must be rigorously separated from residential areas. Telecommunication-based work has few of these undesirable effects, and so affords the possibility of interweaving living and working spaces in a much finer-grained way—a matter of floor plans rather than land use maps.

In other words, the standard land use planning strategies of the industrial city must be inverted. At an urban scale, workplaces and homes no longer need to be kept apart in separate zones. Their intermixture should, in fact, be encouraged. But within the live/work dwelling itself, the need for separation reemerges.

■ The Sociology of Wired Dwellings

At the top end of the socioeconomic food chain, in many parts of the world, demand for these highly serviced live/work dwellings will probably be driven by changes in the composition of the workforce. In particular, if we optimistically assume that glass ceilings will be shattered, and that increasing numbers of women will move into demanding, high-level positions, it will be more and more difficult to sustain traditional spatial and temporal distinctions between professional and domestic roles. There will be a growing need for flexibility in the working hours and conditions of those—both women and men—who give care to children and the elderly, or whose work obliges them to interact across time zones. And as baby-boomers age without mandatory retirement to remove them from the workforce, there will be burgeoning demand for settings that facilitate continued part-time work as consultants and contractors.

At the lower end, by contrast, employers are the ones who stand to benefit most directly. The live/work home shifts the responsibility and cost of maintaining workspace from employer to employee, and makes it far more difficult for union organizers and government inspectors to enforce workplace protections. In the extreme, home workspaces can become exploitive home sweatshops.[6]

For good or ill, then, the home will play a stronger role in our lives than ever. And our close, direct, intense relationships with confidantes, lovers, parents, children, siblings, sharers of dinner tables, bathrooms, or beds, and bringers of chicken soup—those that sociologists term our *primary* social relationships—are generally likely to remain face-to-face and domestically based.[7] To be sure, better communication, allied with swift and efficient transportation, provides us with enhanced capacity to sustain established primary relationships at a distance; scattered extended families can hang more closely together, bicoastal romances have a better chance of success, and travelers no longer need feel so out of touch. But the effect of telecommunications, here, is mostly to create interaction penumbrae—diffuse extensions of face-to-face foci rather than replacements for them.

■ Local Attractions Rule

Live/work dwellings scattered everywhere is one imaginable consequence of electronically reducing the need to locate in proximity to workplaces and services. The iconoclastic urbanist Melvin Webber famously pointed out this possibility in the 1960s: "For the first time in history, it might be possible to locate on a mountain top and maintain intimate, real-time and realistic contact with business or other associates. All persons tapped into the global communications net would have ties approximating those used today in a given metropolitan region."[8] It is easy, from this, to conjure up chilling visions of urban dissolution into endless, undifferentiated suburbia.

But why would you choose that particular mountain top? Presumably because of its scenic beauty. If you're not a hermit or a Kaczynski-esque psychopath, mightn't you rather live at the bottom of the ski lift than the top of the mountain? It only takes a moment's reflection to see that locational freedom does not mean locational indifference.

More precisely, the advantages (or disadvantages) of a particular residential site are a compound of its local physical, economic, and cultural attractions together with the costs—including time costs—of gaining access to related destinations and necessary services.[9] People make tradeoffs; they may choose to accept unattractive residential locations in the interest of accessibility to employment, or accept additional travel times and costs as the price of attractive locations. So relying less on immediate adjacency, and being able to maintain far-flung relationships more effectively through efficient transportation and telecommunication, simply mean that local attractions and disadvantages loom larger relative to accessibility. If you can locate anywhere, you will go where it's nice, or where it is culturally stimulating, or perhaps where you get work done more effectively.

We can expect, then, that localities capable of one-upping others through their pleasant climates, spectacular scenery, and attractive recreational opportunities will attract not only holidaymakers but also a new class of permanent residents—those who can work just about anywhere through electronic linkages, and who can afford to buy into the best places. The Aspens, Tellurides, Malibus, Luganos, and Tahitis of the world will tend to attract populations of high-end teleworkers in fields such as finance, software design, and writing for the entertainment industry.[10]

Cities and towns with unique architectural environments and cultural traditions stand to benefit from the new locational freedom in similar fashion. The gorgeous old city of Venice, for example, has been losing population because it has no room for factories and office buildings (the nearest are across the lagoon in Mestre) and the tourist industry cannot generate sufficient economic opportunity to compensate. But its characteristic and irreplicable attractions remain, and it can integrate modern telecommunications infrastructure far more gracefully than it could ever have adapted to the demands of the industrial revolution, so it has an opportunity to attract footloose teleworkers and recast its famous neighborhoods into a revitalized twenty-first-century form. Many historic, treasured, but economically sidelined cities and neighborhoods—from Bath to Savannah—have similar potential.

When it all shakes out, the guiding real estate principle turns out to be this: telecommunications networking can add great value to localities where relatively well-off people would like to live. It can remove constraints that have prevented them from doing so in the past. But it doesn't do much for localities that have no intrinsic attraction. Nor does it help people who find themselves trapped in marginalized, underserved areas and are too poor to move.

■ Renucleation

Since local scenic, social, and cultural attractions are distributed very unevenly in space, there will still be settlements that nucleate around them. The electronic unraveling of traditional imperatives of adjacency may produce certain urban rearrangements—perhaps major ones—but it is very unlikely to result in random scattering and galloping decentralization. We will continue to see a spatial division of labor, within which different localities perform varying specialized roles according to their comparative advantages. Things will still have their places. It will remain possible to describe neighborhoods, cities, regions, and nations in terms of their characteristic clusters of economic activities.

Local attractions and related activity patterns are, of course, very often social constructions—the outcomes of highly contingent historical processes that have concentrated people, institutions, wealth,

physical infrastructure, and buildings at particular locations. It can
certainly be argued that they were not inevitable. But this does not
make them any less real, or necessarily any less durable. Places like
Wall Street, the City of London, Hollywood, Bollywood, and Silicon
Valley will continue to attract those who want to be where the
action is, and who aspire to the status of insiders.

Indeed, the effect of decreasing reliance on adjacency may actually be even greater *centralization* of particular activities on these sorts
of locations. The elites who control the global economy, and benefit
most directly from it, will want to cluster together at vibrant and
attractive locations. Geographic dispersion of enterprises, and concentration of ownership, control, and profit appropriation, can turn
out to be opposite sides of the same coin.

■ Twenty-four-Hour Electronic Neighborhoods

One potential outcome of all this, where zoning and other policies
allow it, is a clustering of the new-style live/work dwelling in twenty-
four-hour neighborhoods that effectively combine local attractions
with global connections. These—not isolated, independent electronic
cottages—will be the really interesting units in the twenty-first-
century urban fabric. And chances are they will take many different
forms.

Some former bedroom suburbs will probably be able to take
advantage of the fact that they are no longer half-empty in the hours
between the morning and evening commutes, and will refocus themselves around newly viable local services such as neighborhood
schools, child and elderly day care centers, business centers, dry
cleaners, sports facilities and health clubs, and coffee shops and
restaurants.[11] Some downtowns may succeed in remaining vital by
attracting greater full-time residential populations together with the
services that these demand, and will cease to empty out after office
working hours. (This may entail converting former office, warehouse,
and light industrial space to residential uses.) And some former recre-

ational communities at sites of scenic and cultural interest will be able to attract permanent teleworker populations.

In an ironic turnabout, some residential colleges and universities will recognize that their ancient patterns of live/work spaces clustered around communal facilities such as laboratories and classrooms are not anachronisms, but appealing templates for the future. These institutions will not fragment into scattered distance-education enterprises, as some have suggested, but instead will differentiate themselves and compete for the best talent by emphasizing intense face-to-face community in congenial surroundings, combined with efficient electronic linkages to a wider world. These silicon towers will simultaneously be both more concentrated and more connected than campuses of the recent past.

■ **Redistributed Secondary Relationships**

In all these cases, the social effect of restructuring living and working arrangements is largely one of redistributing and relocating our *secondary* social relationships—those with people we regularly encounter, and whose names or faces we know, but with whom we are not so closely engaged as in our primary relationships. This includes our relationships with our friends, daily acquaintances, co-workers, and tradespeople. In secondary relationships, as sociologists point out, we mostly involve people in one or another of their particular roles, rather than interact continually with the whole person.

Preindustrial towns and cities relied heavily upon structures of such relationships, of course, and tended to concentrate them locally, within neighborhoods. In cities of the industrial era they remained crucial as well, but they were scattered far more widely throughout the urban fabric; the more mobile urban dwellers formed them in the workplace and at points of contact with organizations and systems that were important in their daily routines. Furthermore, as many commentators have pointed out, the very possibility of urban public life has depended on opportunities for serendipitous formation of secondary relationships across sociocultural boundaries.[12] If you don't have these, you're living in an interest group or an institution, not a real city.

In the emergent twenty-four-hour neighborhoods of the digital electronic era, patterns will be transformed yet again, and the net effect will be complex. Some secondary social relationships will simply be eliminated as electronic systems replace bank tellers, retail clerks, and the like. But others will be regenerated at the neighborhood level, as local life revitalizes; more of the people that you get to know will be nearby residents. And others still will be formed and maintained at a distance through combinations of electronic interaction and occasional face-to-face meetings. We may find, then, that social integration through secondary relationships occurs at both smaller and larger scales than those that have characterized the industrial era. And the opportunities and limitations will be generated by a mix of tangible and electronic places and boundaries.

■ **Revitalized Local Life vs. the Specter of the Dual City**

The electronically enabled shift of activities back to the home, and the formation of twenty-four-hour, pedestrian-scale neighborhoods that are rich in possibilities for local secondary social relationships, potentially produce the conditions for vigorous local community life, for the formation of social and cultural capital in ways that have seemed lost.[13]

Under the most optimistic scenario, these new patterns will re-create what was best about old-style small towns and urban neighborhoods—the qualities that were celebrated by Jane Jacobs in *The Death and Life of Great American Cities*, that have been so determinedly sought in neotraditional vein by the New Urbanists, and that have been pursued by sustainability-oriented modernists such as Richard Rogers.[14] And they may sometimes succeed in generating hot spots of specialized economic and cultural activity, as in the multimedia-oriented loft communities that have grown up in New York's Silicon Alley area and South of Market Street in San Francisco.[15]

Perhaps this is the best resolution of the increasingly strident
debate between the proponents of globalization and the defenders of
local culture and regional identity: administrative and political units
that can function both locally and globally. But as localities adapt,
with varying degrees of success, to the new conditions and demands,
there will be losers as well as winners. Much existing housing stock
will turn out to be ill suited to the integration of workspace. Low-
income communities may attract less investment in new telecommunications infrastructure, and in any case may lack populations with
the education and motivation to take effective advantage of it. Many
suburbs will prove difficult to adapt to twenty-four-hour life. And
many downtowns will lack the buzz needed to attract permanent
residents. These places will experience the downside of the digital
revolution.

In particular, there is an obvious and serious danger that this
reconfiguration of urban patterns will further cluster the affluent
while leaving the poor in places with few good jobs and services.[16]
Today, for example, high-flying Silicon Valley professionals can commute in their air-conditioned cars from gated residential communities to campus workplaces with guards at the entries, scarcely
noticing that they are passing through marginalized, crime-ridden
areas like East Palo Alto. When they do notice, they probably lock
their doors.

Urban areas could well continue to congeal into introverted,
affluent, gated communities intermixed with "black holes" of disin-
vestment, neglect, and poverty—particularly if, as the unrestrained
logic of the market seems to suggest, low-income communities turn
out to be the last to get digital telecommunications infrastructure and
the skills to use it effectively. As Manuel Castells has vividly warned,
we could end up with *dual cities*—urban systems that are "spatially
and socially polarized between high value-making groups and functions on the one hand and devalued social groups and downgraded
spaces on the other hand."[17] Dwindling opportunities for contact
across the borders of more and more discrete units could certainly
cause public life to atrophy, and we could eventually face the explosive combination of decayed and derelict urban areas ringed by the

territories of psychopathic survivalists barricaded in their isolated electronic forts.[18]

For planners and politicians, steering us away from the dual city is a matter of finding policies that generate an acceptable level of social equity. For architects and urban designers, the complementary task is to create urban fabric that provides opportunities for social groups to intersect and overlap rather than remain isolated by distance or defended walls—the laptop at the piazza cafe table instead of the PC in the gated condo.

■ **And We Shall Build . . . ?**

Ultimately it comes down to a basic social and political choice. What will we use the multifaceted and sometimes contradictory affordances of digital technology *for?* Will we employ them—as seems possible— to help revitalize small-scale neighborhoods and to strengthen interconnections and social interactions? Or will they become a means for the affluent elites to flee the problems of the cities and to create isolated, privileged enclaves while leaving the less fortunate to their fates? Though our options certainly are not unconstrained, the outcome isn't technologically predetermined. Nor is it categorically given by existing geographic patterns and the legacies of history.[19]

In creating their homes and neighborhoods, people will find ways to appropriate and transform the technology in diverse fashions, just as they did with electric power and with the telephone.[20] As existing urban areas come to terms with the digital revolution, and as new real estate developments respond to its demands, we are likely to see both the upside and the downside scenarios played out, in different social and geographic contexts, within different public policy frameworks, and as the result of varied entrepreneurial and design efforts.

Most importantly, this engagement will create opportunities for positive design and policy intervention. You can make a difference, as resourceful and idealistic individuals have done in the face of past urban transformations.

Where will we get together?

What sorts of meeting places, forums, and markets will emerge in the electronically mediated world? What will be the twenty-first-century equivalents of the gathering at the well, the water cooler, the Greek agora, the Roman forum, the village green, the town square, Main Street, and the mall?

■ Online Meeting Places

Many of them will be virtual. Friends and families, co-workers, students, and members of communities of practice and interest will increasingly communicate among themselves by means of software that constructs commonly accessible online places.

They will make growing use of electronic mail systems, mailing lists, newsgroups, chat rooms, Web pages, directories and search engines, audioconferencing, videoconferencing, increasingly elaborate, avatar-populated, online virtual worlds, and software-mediated environments that we cannot even imagine yet. Some of these virtual meeting places will be the private domains of well-defined special groups, some will be discreetly out of the public eye, and some will even be determinedly clandestine; others will be true public space—in principle, at least, open to all.

Whereas physical meeting places depend for their success on centrality within densely populated areas, virtual venues need not. A traditional auction house, for example, is a conveniently located place where buyers and sellers meet, at specified times, to negotiate prices and execute transactions; participation in auctions is limited by accessibility. But an online auction site such as eBay.com connects widely scattered buyers and sellers who would never otherwise have a chance to encounter one another, is as available to Maine villagers and rural Texans as it is to Manhattanites, and operates continuously and asynchronously.

As development of the implementation technologies has progressively loosened constraints, designers of these virtual meeting places have experimented with a variety of formats. And in doing so, they

have raised some fundamental questions. When do we need communication to be synchronous, and when should it be asynchronous?
When should we use voice, and when should we rely on text? When
is it appropriate to remain anonymous, and when should participants
be required to identify themselves? When do simple handles suffice,
and when do we need more elaborate avatars or video self-represen-
tations? When should interaction unfold in a one-dimensional
sequence like the text of a play, when should protagonists occupy
a two-dimensional surface like a frame of a comic strip, and when
should avatars move around in three-dimensional settings?

What images and precedents should guide design? Should two-
dimensional and three-dimensional virtual settings look like places in
the physical world, or—in a domain without materiality, gravity, or
weather—should they look entirely different?[1]

And most importantly, perhaps, who will pay for them, who
will control them, and who will have access to them? Will they be
universally accessible public property, like the streets of a city? Will
they be commercially operated pseudo-public places, like malls and
Disneylands? Or will they be like private clubs, with the electronic
equivalents of velvet ropes and beady-eyed bouncers?

■ On the Line versus Online

Experience has shown, however, that putting your thoughts online is
not the same as putting your body on the line in places like the
Roman forum, Hyde Park Corner, Tiananmen Square, or the Venice
Beach Boardwalk. This has both advantages and dangers.

Most obviously, online meeting places insulate you from physical
risk. You cannot be beaten up by those who take violent exception to
your views. There are no muggers, and no cops with billy clubs. You
will not be confronted face to face by aggressive panhandlers, or by
the mentally ill. This sometimes creates the ground for positive interactions that would not occur otherwise; in Santa Monica, California,
for example, the PEN civic network—which is accessible both from

private homes and offices and from kiosks in public places—has provided a congenial, nonthreatening place for the homeless population and their more fortunate fellow citizens to open up a dialogue. Instead of cruising the personals in the *New York Review of Books* or the *Boston Phoenix*, adventurous lonely hearts can take their chances with jailbabes.com—a pen pal and singles introduction service for women "confined in prisons and correctional institutions all over the country." Even more dramatically, citizens of mutually hostile nations, who have no place to meet in physical space, can often find neutral ground in cyberspace.

Furthermore, you are not compelled to display the usual markers of age, gender, and race. You can hide behind your handle or avatar, and you can readily construct disguises and play roles. So, many online hangouts are like masked balls or Mardi Gras celebrations; they provide well-bounded, socially useful opportunities to experiment with self-representation and alternative identities, and to step temporarily into the shoes of others.

But these liberating affordances can also be put to less desirable uses. Anonymity, and the lowered likelihood of retribution, can encourage ranting and flaming. Loudmouths can blather on endlessly from cyber-soapboxes. And disguises can cloak con men and predators.

So it is far too simplistic to think of online meeting places as direct substitutes for physical ones. Instead, we should treat them as useful new additions to the architect's and urban designer's repertoires—with strengths and weaknesses that fit them to certain purposes but not to others.

■ **A Shift in Scale**

Whatever their norms and forms—and these will probably remain highly varied—online meeting places will allow circles of *indirect* social relationships to widen.[2] Most of these indirect relationships will be tertiary in character—with corporations and bureaucracies rather than particular persons you can name. (When you purchase a volume from an online bookstore, for example, you do not get to know anyone personally, but you do become economically linked to the anonymous employees of that enterprise.)

In other words, you will be able to keep in some sort of contact with many more people, and these people will be spread over wider areas. According to Michael Dertouzos's arithmetic, you could quickly reach maybe a couple of hundred people, back in the days of the village, by walking. The automobile jumped that by a factor of a thousand. Now, computer networks push it up a thousandfold once more—to somewhere around two hundred million.[3] You can quibble about the exact numbers, but the orders of magnitude are surely correct.

In this context, you cannot rely—as inhabitants of small towns and neighborhoods traditionally have—upon repeated face-to-face contact to establish the trust on which intellectual and commercial life depends. Nor do you have the benefit of familiar architectural cues; the dignified stone facade of the local branch bank, for example, with its comforting intimations of solidity, permanence, and reliability, is replaced by the interface of an online home banking or financial management system. So, as Internet marketeers quickly figured out, trusted brand names and brokers play an increasingly crucial role. For organizations with goods and services to offer, maintaining brand equity on the information superhighway serves essentially the same purpose—in a much larger context—as maintaining conspicuous premises on Main Street.[4]

Digital telecommunication thus extends and intensifies the earlier effects of transportation networks, mail systems, the telegraph, and the telephone. It serves as a mechanism for economic and social integration on a large geographic scale, cutting across traditional political borders. It proliferates tertiary social relationships, and the associated mechanisms of branding and broking. And Manuel Castells has suggested that it may also be "a powerful medium to reinforce the social cohesion of the cosmopolitan elite, providing material support to the meaning of a global culture, from the chic of email addresses to the rapid circulation of fashionable messages."[5]

All this would have astonished grumpy old Thoreau, who—rooted in a nineteenth-century conception of local community—wrote in 1854: "We are in great haste to construct a magnetic telegraph from

Maine to Texas; but Maine and Texas, it may be, have nothing to communicate."[6] We know now that they do, indeed, have plenty.

■ Invisible Boundaries

Paradoxically, though, this globalizing effect is accompanied by the creation of new, less visible boundaries. To see why, let us put Dertouzos's
figures in perspective. If you live to a good age, you have maybe half
a million waking hours. If your world of interaction is at a village
scale, each member of it gets, on average, a couple of thousand hours
of your time. At an automobile scale, it is down to two hours each.
And at a global computer network scale, it is reduced to less than ten
seconds. Clearly, then, attention becomes a scarce resource, and intervening attention management mechanisms are essential if we are not
to be overwhelmed by the sheer scale at which electronically mediated global society is beginning to operate.

Mailing lists, newsgroups, personalized news services, information
filters of various kinds, software agents, and other arrangements for
sustaining and managing online relationships play this crucial role.
Reasonably enough, they typically provide efficient means for linking
up like-minded people rather than for confronting differences.
Advertisers, political activists, and others with messages to get out
welcome them, of course, because they effectively segment audiences
and markets.[7] Thus they tend to reinforce sociocultural boundaries
and categorical identities—as professionals in specialist scholarly areas,
members of religious sects, sharers of sexual identities, promoters of
political causes, sufferers from specific diseases, cocker spaniel owners,
Linux hackers, frequent fliers, Buick dealers, cigar smokers, Trekkies,
Barbie doll collectors, or whatever.[8]

It is far too facile, then, simply to equate communication with
community (despite the fact that the terms have the same Latin root)
and to conceive of cyberspace as some sort of vast village green in
the sky. The effects of online interaction are various, complicated,
and sometimes socially and culturally contradictory. While they are
breaking down some established categories and boundaries, online

meeting places can simultaneously strengthen others, and even create new ones. And they are clearly creating a condition under which individuals position themselves less as members of discrete, well-bounded civic formations and more as intersection points of multiple, spatially diffuse, categorical communities.

■ The Virtual Complements the Physical

Of course, time spent interacting online is time spent not doing something else. It is easy to leap from observing this to the conclusion that surfing cyberspace substitutes for more socially desirable face-to-face interaction with family, neighbors, friends, and urban strangers in public places—a chestnut that has routinely been tossed out by recovering netheads, OD'd screen-starers, and computer-jaded curmudgeons.[9] They picture us all huddled at home in our underwear, typing email messages to one another. Under this neo-Durkheimian scenario, anomie rules as never before.[10]

But this reasoning depends on the questionable assumption that our capacities for social interaction are fixed, and thus set up zero-sum games for us; if you devote your attention to certain social opportunities, you must correspondingly decrease your attention to others. There is growing evidence, however, that electronic telecommunication both increases our overall capacity for social interaction and changes the structure of the game in complex ways. The consequences are far from straightforward.

It seems, for example, that so-called "virtual communities" work best when they are allied with the possibility of occasional face-to-face encounters, and that online interaction actually stimulates demand for more familiar sorts of meetings and meeting places. In his lively account of the early online community the Well, Howard Rheingold observed: "The WELL felt like an authentic community to me from the start because it was grounded in my everyday physical world. WELLites who don't live within driving distance of the San Francisco Bay area are constrained in their ability to participate in the local networks of face-to-face acquaintances. By now I've attended real-

life WELL marriages, WELL births, and even a WELL funeral."[11] And
Stacy Horn, founder of New York's Echo, has similarly suggested:
"If someone you talk to online is at all interesting, you want to meet
them. It isn't so much what they look like, you simply want to be with
them *in the flesh*. I don't just want to talk about movies with people, I
want to go to movies with people."[12]

In a broader context, the growth in telecommunications during
the 1980s and 1990s has—seemingly paradoxically—been accompanied by burgeoning demand for hotel meeting facilities and con-
vention centers. Some of this, no doubt, has simply been due to
general economic expansion. But much of it results from a characteristic behavior of geographically distributed businesses, professional
organizations, and interest groups; they form and sustain themselves
by means of electronic telecommunication, then they find that they
need annual face-to-face get-togethers to refresh relationships among
members and to reestablish trust and confidence. And conversely, face-
to-face contacts at these meetings stimulate subsequent telecommunication. The two are inextricably intertwined.

Comparison of telecommunication and transportation demand
statistics tells a similar tale. Generally, the two track in parallel.[13] Unsur-
prisingly, if you make a lot of long-distance calls, you are also likely
to fly to quite a few face-to-face meetings. You can get a lot of bandwidth, when you really need it, by transporting heads attached to
shoulders.

■ Connectivity and Sociability

These interactions of virtual and physical meeting places unfold differently when electronic connectivity is scarce and when it is abundant.
And the locations of connection points matter.

When MIT created its pioneering Athena computer network,
for example, workstations were few and expensive, and for security
and ease of maintenance they were grouped at locations called
"Athena Clusters." These soon became important centers of socializa-
tion among students, not because they were specially attractive places

to hang out (far from it!), and not because students had nowhere else to go, but because they were points of availability of a scarce resource. They functioned much like the village wells of old. Then, when connectivity became far more widely available, their social role began correspondingly to fade.

Similarly, Internet cafes, which provided workstations and refreshments in a convivial setting, experienced a brief burst of popularity when the Internet and the World Wide Web were growing rapidly in popularity but home and office connections were still unusual. They had the additional advantage that working at the computer (like reading a newspaper in more traditional cafes) provided an ostensible reason for spending time in a public place, while observing the passing scene and finding opportunities to meet people. As connectivity became more commonplace, these establishments typically sought to retain their clientele by providing faster connections and machines, unusual and costly types of devices that few would own themselves, and specialized knowledge. And they continued to provide a service to computer-savvy young budget travelers, who used them as an inexpensive means to remain in email contact.

In developing countries (and in poorer areas of developed ones), where the development of high-speed telecommunications infrastructure is likely to lag and where few can afford their own connections and equipment, such public points of access are likely to retain their magnetism for a much longer time. In particular, networks of small, Internet-linked local libraries—along the lines of the Brazilian city of Curitiba's famed Lighthouses of Knowledge—seem particularly promising devices not only for delivering a valuable service, but also for promoting positive social interaction.[14]

Where opportunities for connectivity are abundant, the locations of these opportunities may still be socially significant. If a university simply wires dormitory rooms, for example, it will almost certainly encourage students to stay in their rooms working at their computers, reduce general social interaction, and raise the incidence of conflicts among roommates. But if it goes for laptops rather than desktop devices, provides lots of connection points and power outlets in social spaces and library reading rooms, and implements a dynamic network addressing scheme that allows plug-and-play work anywhere, it will promote mobility among different hangouts, chance encounters, and informal grouping.[15]

■ The Role of Electronic Coordination

Even the most familiar sorts of face-to-face meeting places are beginning to work in new ways, and to rely on the complementary functions of telecommunication. This is mostly a matter of changing scales and schedules.

In the past, meetings often took place without explicit pre-arrangement. The small scales of communities, and the regular rhythms of daily life, assured that you could just show up at the well at the usual time, promenade around the piazza, or stroll down Main Street and be pretty sure of running into people you wanted to see. In large, complex, dispersed cities like Los Angeles, however, the probability of such random meetings is very much lower, so you need to call or email ahead to set a time and a place. And the infinitesimal probability of random face-to-face meetings is a defining characteristic of geographically dispersed, electronically mediated virtual communities. So electronic linkages and associated software will take over from traditional mechanisms in these contexts, playing an increasingly important role in coordinating schedules and arranging meetings. In other words, we will employ quick, convenient, inexpensive electronic telecommunication to make the best use of our relatively scarce and precious opportunities for face-to-face interaction.

You can already test this observation against your own experience. What is the single most frequent topic of your incoming and outgoing electronic mail messages? I'll bet that it's arranging times and places for face-to-face meetings.

It turns out, in general, that telecommunications services and virtual meeting places vastly widen one's circle of active contacts, and some percentage of these then translate into face-to-face meetings. We are not really on the threshold of what Melvin Webber famously —and to city-lovers, hackle-raisingly—called "community without propinquity."[16] (The phrase was prescient, but exaggerated.) Instead, we are seeing the emergence of loosely coupled communities in which physical and virtual meeting places are codependent, coordination is electronic, and a little propinquity goes a long way.

■ Contested Cyberturf

Meeting places have often been contested territory, of course—sites of struggle between those who would preserve privileged exclusivity and those who seek wider and more equitable access, between defenders of various rights and liberties and would-be proscribers of practices that they deem offensive or threatening, between defenders of the status quo and those who would overturn it. Electronically mediated meeting places will be no exception, and indeed are beginning to intensify the debates and struggles by offering some unprecedented extremes.

One possible future of the Internet, for example, is as a vast, worldwide zone of unimpeded interpersonal contact and unregulated free speech—discomfiting local jurisdictions with interests in enforcing their own, narrower norms and standards. Conversely, firewall, encryption, and virtual private network technologies now offer the possibility of constructing impregnable electronic havens not only for those who have legitimate privacy needs, but also for mobsters, tax evaders, kiddie-porn pushers, junk bond moguls, terrorist bombers, drug dealers, and all the other objects of intense interest to three-letter federal agencies.[17] Depending on where you sit and how you see it, digital networks can deliver too much access to other members of society or too little.

More subtly, increased use of telecommunications to arrange and coordinate face-to-face meetings can even further diminish the frequency of urban chance encounters. Once, when you wanted to meet *someone*, you went to places where you could find *anyone*—the piazza, Main Street, the local pub, or even the mall.[18] Now, by telephoning or emailing ahead to arrange precise times and places, you can end up meeting only those you explicitly *choose* to meet. It is efficient, but also a condition that threatens us with loss of public life and growing social fragmentation.

At the extreme, electronic management of face-to-face meetings can render some members of society literally invisible to others. If you don't want to encounter other races, classes, or genders, electronic interaction can effectively make sure that you never have to. You can begin to think that everyone is just like you. This effect is not entirely new—the Greek agora excluded and occluded large parts of the populace too—but, the available means to that potential end are now more powerful than ever before.[19]

■ E-Vox Populi

In the particular case of meetings for political rather than business or social purposes, this fragmentation, specialization, and decentralization of face-to-face interaction sites has potentially far-reaching consequences. Scales and strategies of political organization change.

Traditionally, political power has been exerted, made visible, and architecturally celebrated through physical *assemblies* of kings and courtiers, senates, parliaments, cabinets, councils, and so on. Conversely, if you wanted to overthrow established political power, you assembled "the people" in an urban public place, set up barricades, and marched on the local equivalent of the Hôtel de Ville. If the authorities had the wit and the will, they would try to take the usual countermeasures—dispersion of crowds, prohibition of assemblies, and exile of agitators.

This still happens—witness Tiananmen Square in 1989, or, more happily, Wenceslas Square in that same year—but government no longer has to be so spatially condensed (as it was, for instance, in the Paris of 1848), and political mobilization through the Internet has become both possible and effective. Visibility no longer depends on the physical presence of crowds. When the Zapatista rebels arose in Chiapas in 1994, for example, they targeted not only the Mexican state but also world public opinion; they got their message out electronically, and they mobilized support groups worldwide through the Internet.[20]

The modern dictator's strategy of denying electronic visibility by shutting down radio and television broadcasts can be countered in similar fashion. In 1996, when Slobodan Milosevic silenced Belgrade's pro-democracy Radio B92, it immediately began to generate international pressure by pumping its programming out through the Internet, and eventually forced Milosevic to back down.[21]

Thus Tocqueville's famous insistence on the importance of free political associations, and on the "power of meeting" in forming and sustaining such associations, takes on new meaning.[22] Now, the necessary venues can be found not only in physical space but also in cyberspace, and this opens up fresh, highly effective avenues for political organization and action.[23]

■ Civitas and Urbs Decoupled

We have come a long way, then, from the discrete city-state, with its agora or forum at the center and outer walls unambiguously defining its limits—the sort of arrangement that is implicit in the idea of *urbs* (the territory of the civic formation, such as the seven hills of Rome), as distinguished from *civitas* (families or tribes joined together because they shared the same religious beliefs, social organization, and modes of production).[24]

Now, the boundaries and indeed the very definitions of established large-scale civic units—cities, metropolitan regions, and even nation-states—are being contested at many levels. There is a double threat. On the one hand, global information flows are reducing the importance of old political borders and diminishing the effectiveness of physical public space in producing and representing internal social integration. Simultaneously, electronic privacy and interaction management technologies are creating the possibility of new schisms and subdivisions. We do not have to believe pop-apocalyptic prophecies of the imminent collapse of civic structures and the rise of the sover-

eign individual,[25] but we certainly must recognize the growing slip-
page of *civitas* and *urbs*, and the accelerating breakdown of the old
Oxford Dictionary definition of a community as a "body of people
living in one place, district, or country."[26]

As a result, traditional congruencies of citizenship, public space,
and spectacle—long vital in the functioning of cities—have been dis-
located.[27] The streets and squares of the Renaissance ceremonial city,
for example, were sites for civic and religious processions and performances that were generally attended by the populace on special days.
Alberti could thus speak of the city as a place where one "learns to
be a citizen." But you had to be there. Today, by contrast, we learn to
be citizens of multiple, dispersed, overlapping communities through
diverse electronically mediated means—by surfing into virtual public
places, by participating in electronically arranged get-togethers at far-
flung rented locations, and by watching transmissions from physical
public places that have become—like Times Square on New Year's
Eve—global media stages.

■   Reinventing Public Space

The twenty-first century will still need agoras—maybe more than
ever. But these will not always be physical places. They will operate
at an extraordinary range of scales, from the intimately local to the
global. And even where they *look* familiar, they will no longer function in the same sorts of ways as the great public places of the past.

Under these new conditions, though, the simple, ancient principles of public space remain crucial. If public life is not to disintegrate,
communities must still find ways to provide, pay for, and maintain
places of assembly and interaction for their members—whether these
places are virtual, physical, or some new and complex combination of
the two. And if these places are to serve their purposes effectively,
they must allow both freedom of access and freedom of expression.

Where, among the restructured, diffused, overlapping community configurations of the twenty-first century, will production, distribution, and consumption take place? Where will the enterprises and jobs be? Not only—it seems increasingly clear—at the sorts of sites that have attracted economic activities in the past.

Goods and services flow in new ways in an electronically networked and mediated world, one in which the traditional generators of wealth—land, labor, and capital—are joined and sometimes transcended by fast-flowing *information*. More flexible forms of production, marketing, and distribution emerge, eventually eliminating many traditional constraints on location of commerce and industry and enabling formation of new spatial patterns.

For individuals in their everyday lives, the effects manifest themselves as economic pressures that determine where they can find work, where it's expensive or cheap to live, and where they can most conveniently and effectively gain access to the resources, facilities, and services that they need. For architects, developers, and planners, they show up in changing demands for facilities by type and location, and in shifting opportunities to provide communities with employment opportunities and services. And for civic leaders, they present themselves as questions about how to sustain investments in infrastructure and social services under new rules of this old game.

■ Exchanging Intangible Products

How will these effects first be felt?

Large-scale networks, online transactions, and systems of electronic commerce have most obvious competitive advantages in the case of buying and selling intangible products such as insurance policies. They also come out winners in contexts where they can replace traditional media of exchange such as printed airline tickets.

Purchasing life insurance by authorizing an electronic funds transfer is not, for example, like buying a rug by handing over some gold coins to the dealer; since nothing material needs to be transferred, the whole transaction can swiftly and effectively be accomplished online. What actually happens, in the end, is that databases residing on servers somewhere—at arbitrary locations—get updated to reflect the new balances, relationships, and obligations that follow from completion of the transaction. You don't have to be anyplace special to participate; you just need to be connected. It is quick, cheap, and convenient for all concerned.

The process of finding what you need in the markets for these intangibles is different, too. From a consumer's viewpoint, nothing now beats online shopping for the cheapest airline ticket to a particular city on a particular date, or for the most competitive available mortgage rates. But it is the international currency market that provides the most vivid example of the new, high-speed, staggeringly high-volume global traffic in abstractions. Once, currencies were simply precious physical commodities—gold because it was scarce and compact, rum in the early Australian colonies because it was one of the few things valued by everyone, and heavy iron bars in ancient Sparta to discourage complex commercial transactions and focus attention on more macho martial pursuits. Then there were pieces of paper, bookkeeping entries, and bank accounts that represented such commodities. The direct connection to physical commodities gradually weakened, and was finally severed in 1971 when Richard Nixon ended the convertibility of the dollar into gold; the age of floating exchange rates had begun in earnest.

Meanwhile, the telegraph, the telephone, and the telex had begun to connect currency trading rooms around the world, and a relatively high-speed (but still low-volume) international currency market had begun to form. Then came computers and networks, and by the early 1990s the banker Walter Wriston could matter-of-factly write: "The new world financial market is not a geographic location to be found on a map but, rather, more than two hundred thousand electronic monitors in trading rooms all over the world that are linked together."[1] Twenty-four/seven, they set bulls and bears speed-skating around the globe.

Stock exchanges have followed a similar path.[2] Before telecommunications, they were local, face-to-face affairs; the United States had

250 of them in 1850. By 1900, the telegraph and the tickertape had allowed the New York Stock Exchange to emerge as a dominant national trading center. And as 2000 approached, newer exchanges like Nasdaq were taking the form of omnipresent digital electronic systems rather than buildings in particular cities, online brokers such as E★Trade and DLJdirect were offering Internet-based service anywhere in the world, and many grand old exchange buildings—such as the Palais de la Bourse in Paris—had literally become museums. The trading floor of the New York Stock Exchange still buzzed with activity until the closing bell, but cyberspace had stealthily displaced Wall Street as the new, twenty-four-hour capital of capital.

In general, markets have dramatically dematerialized. In the guild-based medieval city, "market" referred to an identifiable physical place where actual goods were exchanged, as in "to market, to market, to buy a fat pig—home again, home again, jig-a-jig-jig." By the time of Adam Smith, the term had begun to designate abstract, spatially ambiguous systems of information and exchange that were more usefully described by the equations of economists than the drawings of architects. (Where else could invisible hands operate?) And when the New York stock market teetered and plunged in August 1987, it was not the building that collapsed. What happened was a sudden, sweeping transformation of economic relationships propagated worldwide, at high speed, by telecommunications networks and computer software.[3]

■ Delivering Information Products

But it's not just abstract quantities like currency and the increasingly complicated financial instruments that have been cooked up in the computer era. Where sufficient bandwidth is available, familiar sorts of information products can be separated from their traditional material substrates and distributed widely and inexpensively via computer networks.

Instead of printing, warehousing, and mailing a technical journal, for example, you can mount the same text on a Web site. With faster servers and higher network speeds you can do the same with photographic libraries and audio recordings; by 1999, record labels were beginning to distribute music online, and the *New York Times* was proclaiming that "the record store of tomorrow will be the desktop computer or digital listening device."[4] And with higher capacity still, you can replace videotapes and video rental stores with digital video-on-demand delivered directly to homes and businesses. All this has, of course, led to intense competition among giant international news and entertainment empires for control of the means of electronic distribution—telephone wires, cable networks, wireless channels, and communications satellites.

Even more dramatically, computer software no longer has to be delivered on floppy disk, CD, or tape; it can simply be downloaded over a computer network. This has yielded various forms of dispersal of the software industry. Some networked-linked enterprises at sites of inexpensive labor engage in bodyshopping—obtaining software production contracts from distant clients, then hiring local contract workforces to fulfill them. Others put together skilled professional teams at attractive locations, then seek software research and development projects from clients worldwide. Yet others produce consumer software products and distribute them to geographically dispersed customers.

Sometimes, the combination of rapid electronic delivery with convenient time zone differences allows an effective new form of twenty-four-hour shift work. International architectural and engineering design firms can, for example, establish offices in cities approximately eight hours apart, then electronically hand off CAD files from one to the other in a continual circle around the globe. Such systems can sometimes be organized to take advantage of particular local capabilities. Thus London's Soho—a hotbed of film and video post-production talent—finds itself in the fortunate position of being about a half-day displaced from Hollywood; it can electronically receive footage at the end of a day's shooting in California, do the post-production on it during the normal London working day, then send it back before the start of the next day's shooting.

In all such cases, where pure information is, in itself, the valuable thing, there seems little doubt that digital network delivery will eventually win out—especially where timeliness matters. Older means have about as much chance as the horse against the internal combustion engine.

■ Remaking Making

With other types of products, those that retain a material component, the availability of digital networks opens up the possibility of radically decentralizing physical production—a surprising inversion of the taken-for-granted centralizing tendencies of the industrial revolution.

Consider newspapers, for instance—products that have traditionally been printed at large central plants, then distributed through elaborate transportation networks. Under this centralized mass production system, everyone gets exactly the same thing. In the early days of telecommunications it became feasible, instead, to transfer page layouts electronically to collections of regional printing plants that were closer to customers, then to add some local content, and so to create regional editions. Today, with the development of pervasive networks and inexpensive home printers, it is increasingly attractive to contemplate personalized newspapers that are printed right at the point of consumption (for those who—like most of us—still prefer their news on paper, rather than on a screen). Once, it made sense to print, then distribute; now it may be better to distribute, then print.

Even the traditional-looking book that you are holding in your hand right now—an artifact that Aldus Manutius would have had no difficulty recognizing—is, in fact, a digitally mediated product. You may have picked it up from a traditional bookstore, or you may have purchased it from an online bookstore. In the latter case, you surfed into a Web site, located the title in an online catalog, filled in an on-screen form to place your order, and received delivery through the mail or via package express. Under this newer system, electronic exchange of information replaces the face-to-face purchase transaction, warehouse and retail space end up at very different locations, and direct, personalized delivery from warehouse to consumer substitutes for bulk transfer of items to an intermediate storage point.

And even if you bought this book from an old-fashioned store, your retailer probably used an electronic purchasing system to order it from the publisher and an electronic inventory control system to keep track of it. Furthermore, if you were to go back down the supply chain and examine the relationships among the geographically distributed team of author and author's assistants, editors, designers, paper suppliers, printers, binders, warehouse workers and managers, shippers, and publishers who combined forces to produce this artifact, you would find widespread and increasing use of EDI (electronic data interchange) to coordinate the dispersed production process and speed it along. The same holds for pretty much any modern product that you can imagine.[5]

Forget those old images of Charlie Chaplin and Lucille Ball struggling with inexorable industrial production lines, anxiously aware that their supervisors were not far away. Such lines still exist, of course, but they are now only a small part of the story. Standing behind each one is a vast, scattered web of remotely coordinated international connections and flows.

■ Value from Knowledge

This story becomes even more dramatic when we explore the sources of a modern tangible product's value and compare their relative magnitudes.[6] In the case of this printed volume, for example, little of the value is in the raw materials and much is in the writing and the design—tasks that could have been carried out almost anywhere, and which yielded readily transferable digital files.

This applies, as well, to products that we do not usually regard as containers of information. In a silicon chip, only a couple of percent is in the cost of the raw materials (mostly just sand, after all), and most of the rest of the value is added by the extraordinarily intricate design and the translation of that design into instructions that drive computer-controlled machinery. Even in the most traditional of industrial products, such as steel girders, an increasing percentage of the value derives from information processes that are not tightly tied to particular industrial sites.

In general, the relative contribution of knowledge to a product's value is rising. And so, simultaneously, is the possibility of supplying that knowledge from afar.

■ Relocating Production

The effects of this vast transformation of product design, ordering, and delivery on business and industry location, on the organization of transportation systems, and ultimately on employment opportunities at specific locations, are profound.[7] The large-scale, complex material-processing systems that characterize modern industry are coordinated and controlled in startling new ways, and end up being distributed in new spatial patterns.

While some production facilities still need to locate in close proximity to sources of energy and raw materials (as in the cities of the industrial revolution), many now depend far more heavily on coordinated use of fast, flexible telecommunication and transportation networks to link them to their widely distributed technicians, suppliers, and partners. Some great manufacturing centers, such as Hong Kong, are now less sites of actual factories than command-and-control centers for geographically distributed value networks. And in forming the links in these networks, software compatibility may be much more important than propinquity.

Fittingly, the semiconductor industry provides one of the most dramatic examples. Representing the most traditional organizational structure, there are companies that both design chips and produce them in their own fabrication plants—though there is little need for

the design and fabrication facilities to be near one another. Then
there are chip foundries that produce chips to others' designs. And
finally, there are companies that design, market, and distribute chips
but have no fabs of their own, and hire production facilities—which
might be pretty much anywhere—as they need them.

■   **Make after Buying**

Telecommunication can also enable more direct and immediate connection of producers to their customers, so reducing or eliminating
the roles of local dealers and other such intermediaries, and significantly cutting inventory costs.

In 1996, for example, Dell Computer Corporation launched
www.dell.com—a Web site for direct purchase of computers.
Customers could surf in from pretty much anywhere in the world,
configure a computer online, then send the order to the manufacturing plant, where the specified machine was assembled and
shipped within hours. Within a couple of years, Dell's competitors
were scrambling to catch up.

And consider something as homely as a pair of jeans. Once, you
had two options. You could buy an inexpensive, standard-sized pair
that was centrally mass-produced, warehoused by a wholesaler, then
shipped in bulk to your local store; that, of course, is the familiar
industrial way. Alternatively, in a throwback to an earlier era, you
could have yourself measured for a very much more expensive pair
from a tailor—the local craft way, and still a vigorous Hong Kong
tradition. But in 1994, Levi Strauss radically reconfigured the production and distribution system by putting computerized measurement-
taking systems into stores, electronically transferring custom orders to
the factory, laser-cutting and bar-coding the pieces, stitching them on
the regular assembly line, and finally mailing the finished products
directly to customer addresses.[8]

In these cases, it's back to make-after-buying, rather than buy-
after-making—but with a new, postindustrial twist.

■ The Recombinant Workplace

Not only are the locations of work sites changing, so are their characters. Familiar types of workplaces are fragmenting and recombining into new patterns.

As Ithiel de Sola Pool has observed, this also occurred in earlier eras, with the coming of the telegraph and the telephone. Courtesy of the telephone, "corporate offices moved away from the factory, which could be adequately controlled by a phone call to the hired manager; the president moved downtown, where he could have face-to-face meetings with bankers, suppliers, and customers." Consequently, downtown changed "from a collection of specialized neighborhoods to a dense concentration of business offices engaged in commerce with one another."[9]

At production sites today, network-enabled remote monitoring and control means that there is even less need for early-industrial-style machine minding. This function migrates to centralized control rooms, which are not necessarily in very close proximity to the facilities they supervise and which require fewer staff.

In retailing, the elements of a traditional store—sales floor, stock room, and backroom administrative space—may blow apart when electronic interconnections are introduced. The sales floor may be replaced by a system of remote service sites that maintain online catalogs and answer email and the phones, or by small look-and-order showrooms in high-traffic locations such as airports. The stockroom may become a big-box, centralized warehouse and distribution center located near a package delivery hub. And the billing, ordering, and other administrative functions may be performed by teleworkers at home or in nearby telework centers.

And in offices, electronic interconnection dissolves the traditionally tight spatial relationships among private workspaces such as office cubicles, group workspaces such as meeting rooms, informal social spaces, and resources such as files and copying machines. When files are online, and office workers have personal computers and printers, there is no longer much need to cluster private workspaces around central resources; these spaces can migrate to the home or to satellite locations, they can follow employees on the road, or they can transform into "hot cubicles" that are not permanently allocated to particular employees but are reserved and occupied as needed. Meeting rooms and informal social spaces are augmented by virtual meeting places and groupware, but the need for face-to-face meeting space remains; indeed, it may end up becoming the stable nucleus of office workplaces that are far more fluid than those of the past and look more like clubs or hotels than cubicle clusters. The individual components of office work may be mobilized and dispersed, but the group components are likely to remain more place-specific.

All this is enough to challenge the very idea of a business or industrial firm. In his much-quoted analysis of why firms exist at all, Ronald Coase suggested in the 1930s that they created relatively efficient internal information flows, and so minimized the costs of transactions and of the information that workers needed in order to play their roles effectively.[10] Traditionally, a large part of this efficiency has derived from putting workers together under one roof, where they can talk to one another and pass paper back and forth. But as many business commentators have been quick to notice, networks and smart spaces are greatly reducing transaction costs among more ad hoc, geographically scattered groups of collaborators, and so making these sorts of nontraditional units increasingly competitive.[11] As this becomes increasingly obvious, and businesses try to figure out what to do about it, we will hear more and more breathless talk of "virtual corporations" and "extended enterprises."[12]

■ Mobilizing Enterprises

The various new sorts of electronic linkages among employees, consultants, suppliers, manufacturers, distributors, and customers, unlike those accomplished through physical proximity, can rapidly be reconfigured in response to changing conditions and competitive pressures.

Globally mobile capital drives this ongoing process of reconfiguration and adaptation by continually seeking out locations where labor markets and general business conditions are currently most attractive, while multinational corporations take vigorous advantage of their ability to distribute their activities in pretty much any ways they may choose. As Lester Thurow has put it, "The global economy simultaneously permits, encourages, and forces companies to move to the lowest-cost locations."[13] Furthermore, since capital can now migrate at far faster rates than people can, multinational capital can effectively use the threat of withdrawal from a community and so can more readily get the upper hand in its dealings with labor and with governments.[14]

Commentators from both the left and the right are remarkably consistent in their analyses of these phenomena—if not in the lessons they draw. In his magisterial work on nations and nationalism, Eric Hobsbawm observes: "City states like Hong Kong and Singapore revive, extraterritorial 'industrial zones' multiply inside technically sovereign nation-states like Hanseatic steelyards, and so do offshore tax havens in otherwise valueless islands whose only function is, precisely, to remove economic transactions from the control of nation-states. The ideology of nations and nationalism is irrelevant to any of these developments."[15] And George Gilder gloats from the opposite wing: "Capital is no longer manacled to machines and places, nations and jurisdictions. . . . Companies can move in weeks. Ambitious men need no longer stand still to be fleeced or exploited by bureaucrats. Geography has become economically trivial."[16]

Communities throughout the world are increasingly feeling the effects of all this. Once, many of them were held in place by relatively stable, long-term relationships of their inhabitants with local banks, producers, and retailers who provided jobs, conducted business among themselves, and supplied most of the necessities of daily life— the sort of close-knit, highly personal commercial and communal structure that was sentimentally celebrated in *It's a Wonderful Life*. Most people had a long-term stake in the character and quality of the local community, and it paid to be public-spirited. But we will not again see the likes of George Bailey and the Bedford Falls Building and Loan, and the electronic workplace doesn't have that Frank Capra glow.

It's not, we should be careful to note, that economic globalization is really such a new phenomenon; many more literate commentators have gleefully pointed out the comically close parallels between rabid globalize-or-die rhetoric and that of Marx and Engels in *The Communist Manifesto*. They have a point; George Bailey would have been quite well aware of economic developments on the other side of the world, and would often have been affected by them. But digital networking increases the flow of the information that binds enterprises together, and allows transactions to tick over at much higher clock speeds. Now we have not only a global economy, but one that responds (and must be responded to) very rapidly, and that threatens old stabilities as a result.

■ New Game in Town

What can we do about all this? How can we pursue the potential benefits of the emerging new order while avoiding the downsides?

Clearly we have to create new sources of urban economic vitality. To thrive, cities have always needed to put together some economically potent, sustainable combination of natural resources and transportation connections with available land, labor, and capital. Now, in the wake of the digital revolution, the rules and payoffs of this ancient game are changing.

In the past, for example, many cities have succeeded by exploiting local natural resources. The famous mill towns of New England grew up around sources of abundant water power. In Australia, Ballarat, Bendigo, Kalgoorlie, and Broken Hill boomed at sites of rich mineral deposits. In the American Southwest, local petroleum drove the growth of Los Angeles, Denver, Houston, and Dallas. Of course, as the later fates of many of these cities demonstrate, the strategy falters when the resources run out, the prices drop, or new technologies spawn effective competitors.

Other cities have capitalized on their strategic locations to become trading centers. Venice and Singapore found themselves astride major international trade routes, and made shrewd use of the fact. Chicago grew as a crucial railroad center. Amsterdam now derives much of its economic vitality from its role as a major air transportation hub.

In the digital era, an increasing number of cities (Palo Alto, California, is one striking example, and India's Bangalore is another) will find that they can succeed in yet another way—by exploiting their unusual *human resources* to attract and retain economic activities that could, in principle, be located just about anywhere.[17] To win at this game in the long run, they will need the right sorts of local attractions to retain the talent—in particular, pleasant and stimulating local environments, high-quality educational and medical services, and sufficiently flexible transportation infrastructures and building stocks to accommodate rapidly reconfiguring patterns of activity.[18]

But all this clearly depends upon effective strategies for sustaining social investment under the condition that geographic communities and economic communities are no longer coextensive in either space or time.[19] How can enterprises with global interests be motivated to support infrastructure construction and maintenance, preservation of environmental quality, and provision of good education and medical care in particular local contexts? How can the notoriously short time horizons of these big-time economic actors be extended far enough to make a real difference? How can they become committed citizens of the diverse and scattered local communities that they engage?

These will be life-or-death policy questions for civic leaders of the twenty-first century.[20] Get the answers wrong, and face the specter of the Schumpeterian dumpster. Get them right, and cities may—as some optimistic commentators have suggested—be "poised for a huge surge in economic growth."[21]

In ancient Rome you got better military protection and better spectacles than in the provinces. In Manhattan you get better medical care, restaurants, and haircuts than in a cow town. As everyone knows, availability of high-quality services is a major attraction of urban areas.

In the emergent computer-networked world, though, this reduces to a half-truism. Some services still depend on the local presence of the providers, but others can effectively be summoned and delivered from a distance. As a result, new patterns of service distribution are superimposing themselves on cities and rapidly displacing some older ones.

■ A Typology of Service Systems

Service systems consist, in their barest and most obvious essentials, of service *providers,* service *consumers,* and effective means to *connect* the two. The various possible patterns of connection define an elementary typology of service systems. And the effects of digital telecommunications vary by type.

Before telecommunications, the privileged surrounded themselves by servants or slaves and summoned them verbally as required. Members of large service staffs were classified and named according to their roles—butlers, chambermaids, cooks, chauffeurs, gamekeepers, viziers, broom-wallahs, personal trainers, scribes, amahs, corporate lawyers, whatever. It mostly depended on maintaining close physical proximity, even when primitive systems of bells and buzzers began to augment direct verbal communication. And it reflected itself architecturally in provision of servants' quarters, janitor's cubbyholes, service stairs, outer offices, and so on.

As large modern cities grew, so did an alternative system of centralized service points—particularly for sophisticated, specialized services. These allowed economies of scale and could serve large populations at relatively low cost, but service consumers had to travel to them. Medical care, education, and many commercial services followed this pattern, and associated building types such as modern hospitals and schools emerged as a result.

One way to resolve the inconvenient contradiction between gaining economies of scale through centralization and remaining close to consumers through decentralization was to develop distributed systems of branches. Thus, for example, in the nineteenth and earlier twentieth centuries, large banking organizations had head offices in prominent downtown locations, back offices for centralized processing activities in low-rent suburban space, and huge numbers of branches to provide customer service in local communities. Retailing followed a similar pattern. The general result was that Main Streets, commercial strips, and shopping malls became collections of branches and franchises. And the downtown office towers in all but the largest global cities mostly contained branches of national and international organizations.

Yet another strategy was to service distributed populations by means of mobile providers. This has its roots in the ancient tradition of itinerant healers, teachers, peddlers, and cops on the beat. The disadvantages are that the mobile providers have to carry their tools of trade with them, and that it is hard to achieve economies of scale in this way.

Finally, in efforts to maximize the advantages and minimize the disadvantages of these basic patterns, all manner of hybrids developed. A large, central medical facility might be combined with a system of local clinics, mobile care units, and home care providers. A retailer might make use both of downtown showrooms and traveling salesmen.

■ Summoning Assistance

In the nineteenth century, early telecommunications technology was quickly adapted to the task of summoning mobile service providers from central locations as required. This speeded response times, and so made centralized services far more effective.

In 1852, for example, Boston began constructing a system of telegraphic call boxes connected to fire stations, and other cities soon followed.[1] Together with the replacement of hand-drawn firefighting apparatus by horse-drawn fire engines, then by motorized fire trucks, this allowed stations to serve wider areas and larger populations.

Successive waves of telecommunication and transportation technology extended and elaborated the idea. By the 1880s, telephones were being installed in police stations, and police forces eventually began to make combined use of the telephone, the two-way radio, and patrol cars to service extensive areas. From 1928, the Royal Flying Doctor Service began providing medical care to the vast, sparsely populated Australian outback through use of light aircraft summoned by means of pedal-powered Morse code radio transmitter-receivers. Now, in the era of cellular telephones and pagers, service providers of every kind—from drain cleaners to brain surgeons—can continually be on call.

■ Keeping Tabs

In all of these systems, somebody still has to take the action of calling the cops, the medics, the firefighters, the plumbers, or the caterers. But by adding sensors to remote summoning capabilities, the tasks of monitoring need and summoning services as required can be automated.

Thus it is now routine to install fire and smoke detectors in buildings; these not only sound a local alarm, but in many cases they automatically call firefighters to the scene. Burglar alarms that sense opening of doors, breakage of windows, or motion within interior spaces function in much the same way. Continuous electronic monitoring, based on embedded sensors, is beginning to revolutionize the maintenance of structures such as bridges and dams. In industry, it has long been commonplace to embed sensors in plant and machinery to signal malfunctions, and in a ubiquitously networked world this idea will increasingly be extended to automobiles and to domestic appliances of all kinds.

Consider automobile and truck tires, for example. Traditionally, it has been up to drivers and service mechanics to check the pressure manually, and to adjust it as required; neglect results in poor performance and excessive wear. A smart vehicle might perform these routine service tasks itself, making use of onboard pressure monitors, computers, and controllable pumps and valves to maintain constant pressure. But even smarter logging trucks in Alaska and British Columbia now link their onboard computers, via satellite, to geographic information systems and weather information systems and dynamically adjust tire pressures to current conditions. Overkill? Not if the performance payoff is there.[2]

What works for structures and machines can also work for our own bodies. We are also likely to see proliferation of sophisticated medical monitoring devices hooked up to health care providers; once available only in hospital beds, they will increasingly take the form of unobtrusive wearable devices and of systems that provide continuous monitoring in the homes of those who require it.

In contexts where automated monitoring is not feasible, or for some reason does not suffice, electronically mediated remote monitoring may be the next best thing. (Where the industrial revolution spawned machine watchers, the digital revolution proliferates screen watchers.) Remote monitoring tasks may be performed wherever the time zone is convenient, the skills are available, and the price is right. Paramedics in Manila might well provide medical monitoring services to retirement communities in Palm Springs, and dispatch local medical practitioners as required. Similarly, remote eyes and ears might monitor security camera screens and call the local police or security service where necessary.

■ Surveillance and Seclusion

All this, of course, superimposes yet another layer of electronically mediated social relationships on daily life. Wherever such electronic monitoring is carried out, it adds what are sometimes called *quaternary* social relationships—those that exist between observed and anonymous observer—to our primary, secondary, and tertiary relationships. And as vigilant civil libertarians have been quick to point out, we could well end up imprisoning ourselves in a vast electronic Panopticon.[3]

More subtly, we will increasingly face tradeoffs between maintaining our privacy and getting better service by giving some of it up. If an online bookstore or CD store keeps track of your purchases, for example, it can automatically compare these with the purchases of other customers and use these comparisons to tell you what customers with similar interests to yours have been buying. It is a very effective collaborative filtering and recommendation mechanism, and it adds considerable value to the bookseller's service—but you might want to opt out of it if you discovered that the purchase profiles were also being sold to direct-mail marketeers.[4] And you might be *very* upset if you discovered that nosy journalists were poking around in them.

What if you were checking into a fancy hotel? If the hotel could electronically access a detailed profile of your needs and preferences, it could organize the space and the menu to your liking. But is it worth it? Would you want to reveal so much in order to gain this benefit? Then, what about checking into a hospital? Does the greater criticality of the situation change things? Would you be prepared to reveal much more about yourself if it made a significant difference in the quality of your health care?

The crucial policy and design issues raised by superimposition of quaternary relationships are ones of forming appropriate, context-dependent balances. Different people, at different times in their lives, in different relationships to society, balance independence and dependence in different ways. They seek and require different combinations of anonymity and identifiability. Sometimes they want seclusion, and sometimes they want to present themselves very visibly in public. Traditionally, it has been possible to shift these sorts of balances by moving from one place to another. Electronic remote monitoring and summoning technologies widen the range of options, change the payoffs and dangers, and require us to rethink our architectural and legal mechanisms for getting the balances right.

Under the most pessimistic scenarios, the mechanisms will inevitably fail, the powerful will always get whatever information they want, and we will end up with no privacy left. Under more optimistic ones, we will find effective ways to treat identity as an electronically metered commodity. We will turn it up or down, depending on the context.

<h2>■ Delivery at a Distance</h2>

Remote monitoring and summoning significantly change service systems—particularly medical care and emergency services—but it is remote delivery that *really* makes a difference. If you can pump a service out through a network, then you can extend the service area to wherever that network reaches—potentially globally. This creates large service markets, promises greater distributional equity, and is particularly good news for inhabitants of remote and underdeveloped areas, and for those who are immobilized by age or infirmity. Furthermore, the service agent on the end of the line can often become an unsleeping piece of software rather than a human operator.

In the most elementary case, as with audio and video entertainment, news, and some educational services, delivery reduces to transmission and display of an information stream. It may be synchronous, as with radio and television broadcasts, or it may be asynchronous, as with Web news servers. Either way, the network simply provides one-way pipes; the logic is little different from that of water supply systems.

With two-way telecommunication, remote delivery becomes an attractive option for service businesses that pursue the strategy of informing clients about their options, advising them on their choices, then eventually executing a purchase transaction of some kind. It works even in cases where the product or service purchased is eventually delivered in an utterly conventional way.

Travel is typical. Once, you had to go to a railway station, steamship office, or local travel agency to get information and advice and to purchase your tickets. Then, when the telephone came along, you could get much the same service by dialing in; airlines and other transportation companies began to rely on their call center operations, and travel agents began to spend most of their time on the phone. More recently, interactive Web sites have offered a third alternative; you can use them to consult comprehensive online databases of travel information, to conduct sophisticated searches to find flights and rates that meet your requirements, then immediately to make reservations and ticket purchases by executing online transactions. This has made it tougher for travel agents to survive (as they traditionally have) on commissions from ticket sales, and has forced them to compete on the quality of the information and advice that they can offer.[5]

Some areas of retailing are going the same way. Online book and CD stores such as Amazon.com do not just offer convenience and twenty-four/seven service; they also compete with traditional bookstores by offering increasingly sophisticated information and advice. Their catalogs are extensive and detailed, contain summaries, reviews, and cross references, and can be searched in a variety of ways. Furthermore, they can build customer loyalty by offering collaborative filtering and recommendation services that become more effective the longer and more consistently you shop there. Thus they are far more formidable than their mail order and phone order predecessors.

Banks and financial services vendors have also been affected dramatically. Deposits, withdrawals, and balance queries have become high-volume, low-price commodity transactions; they are increasingly handled remotely and automatically by ATM machines and electronic home banking systems, rather than by tellers behind counters as in the past. Bills get paid online rather than through the mail.[6] And an increasing number of investors use inexpensive online trading sites instead of calling live brokers.

In this new competitive environment, vendors seek to differentiate themselves by the quality of the online information, analysis, and advice that they can offer. So electronic home banking systems get integrated with personal financial management software. Retirement fund providers create elaborate Web sites incorporating up-to-date statements, educational materials, benefits calculators and other decision-support tools, and online transaction capabilities. Online stock-trading and mutual fund sites provide customized portfolio trackers, real-time stock quotes, moving average charts, quarterly estimates, economic calendars, reports from analysts, and personalized recommendations in place of a broker's counsel.

The scoffers who think these online services are little different from mail-supported and phone-supported services of the past, and that nothing can replace face-to-face interaction with a knowledgeable human specialist, are just not getting it. It is the ubiquity and speed of electronic delivery capabilities, combined with the possibility of effectively integrating electronic intelligence, that makes the crucial difference. Networks are opening up vast markets for both familiar and radically innovative services, entrepreneurs are responding, and a new kind of electronically mediated service economy is rapidly emerging. An increasing number of traditional service businesses will suddenly find themselves Amazoned by upstart dot-coms.

■ Expanding the Web of Indirect Relationships

The general social effect of these new sorts of teleservice systems is to eliminate familiar middlemen. They are replaced by electronic systems and software.

Instead of going to a branch bank and meeting with a teller—perhaps someone you have come to know through regular contact—you interact with a faceless ATM machine or an electronic home banking system. Instead of buying a theater ticket at the box office (or from a scalper), you surf into a Web site, select a seat from an on-screen plan, and prepay by credit card. Rather than visit your friendly local retailer, you run searches on catalogs and click "order" buttons. Where you once had to line up at the Registry of Motor Vehicles to renew your driver's license, you now perform that task online.

So indirect, anonymous, electronically enabled relationships are proliferating in our daily lives, while certain kinds of face-to-face transactions (and the secondary social relationships with familiar intermediaries that these have fostered) are correspondingly being reduced. Society as a whole is becoming more and more dependent on a vast, complex web of automated, electronic intermediation—

our new, all-purpose go-between. The reductions in transaction costs, and the gains in market efficiency, are potentially enormous; not surprisingly, Bill Gates has written lip-smackingly of a coming era of "friction-free capitalism."[7]

Many people, understandably enough, fear reduction to a bit-sucking subspecies of *homo economicus,* and the attendant loss of human contacts and relationships. But let's ask the hard questions. What are these particular social relationships really worth? And what will replace them? For my part, I can well live without the human contact that I used to get from the bored and overworked clerks at the RMV, and I can put the time that I spent in line to much better use. And I doubt that I am alone in this.

The point, surely, is not to leave a void, or to fill that void with game shows and sitcom reruns. If electronically achieved efficiencies are to yield real human benefits, they must be complemented by opportunities to spend freed-up time on something *better*—where "better" is defined in social as well as individual terms. This is a crucial policy and design challenge. We will be able to count the wired, live/work home a social success if it can provide opportunities to devote more time and energy to our most valued primary relationships. The small-scale twenty-four-hour neighborhood will be a winner if it can encourage and reward renewed attention to community building. And perhaps time that would have been wasted in searching out and *buying* books will now be spent more productively in *reading* less expensive and more readily accessible electronic publications.

■ Telerobotics

All this applies to services that can successfully be disembodied. But what about those that traditionally have required not just exchange of information but also human hands, right there, on the spot? Can you get your car fixed remotely?

Well, you can certainly get your computer fixed—under some circumstances, at least. If you let a skilled technician log in remotely, you may be able to get software problems resolved without a site visit or taking the machine to a service center; large-scale networks would, in fact, be very difficult to maintain without this sort of remote servicing. As more and more machines get embedded software and network connections, they will also be serviced in this fashion. If they cannot be fixed directly, their problems will be diagnosed remotely, and technicians dispatched with appropriate parts and tools.

Where this strategy does not suffice, telerobots may, at least in principle, be brought in to do the job. Telerobots are remotely controlled machines that are capable of performing various physical tasks. They may be fixed in place, like industrial robots, or they may be mobile, like delivery vehicles. They may be wired up to telecommunications networks, or they may be linked wirelessly. Their every move may explicitly be controlled, or they may be equipped with some autonomous decision-making capability.

The irony-dripping Telegarden project, created by Ken Goldman and Joseph Santarommano, was an early and provocative exploration of some of the ways in which network telerobotics might work. It was a telerobotically tended garden accessed through the Web. You could join the community that jointly maintained it by providing your email address to the project organizers and to the other gardeners. Membership in this community allowed you remotely to maneuver a robotic arm via a Web interface, to plant and water seeds, to monitor all actions, and to view the state of the garden. *Gardening Design* magazine—not usually noted for its attention to the digital world—was moved to comment: "Sowing a single, unseen and untouched seed thousands of miles away might seem mechanical, but it engenders a Zen-like appreciation for the fundamental act of growing. Though drained of sensory cues, planting that distant seed still stirs anticipation, protectiveness, and nurturing. The unmistakable vibration of the garden pulses and pulls, even through a modem."[8]

On a quick tour of the Web you can find many engaging and amusing telerobotic toys and art installations. (Yahoo classifies them as "Interesting Devices Connected to the Net.") As I wrote this paragraph, I discovered sites that allowed you—or promised to allow you, or once would have allowed you—to excavate artifacts in a sand-filled

terrarium, to control various items of laboratory equipment, to pan
and tilt video cameras at many locations, to operate model trains in
Germany and watch them move, to switch on the lights decking a
faraway Christmas tree, to run several automated telescopes, to move
blocks around with a mechanical arm located at the University of
Western Australia, to paint pictures with real paint and brushes, and
even to make remote toast.

In general, telerobotics sounds complicated, expensive, and a bit
perverse. Indeed it often is, but it can make practical sense under conditions where distances and travel costs are very high, where services
must be delivered to dangerous localities, or where demand is widespread but skilled providers are confined to a few locations. Consider
specialized surgery, for example. No doubt it normally makes eminent
sense to have the surgeon in the room, with the patient. But is it necessarily better to transport a sick patient or busy surgeon over very
long distances if a combination of telerobotics and digital imaging,
delivered via a smart surgical suite, might substitute? What about bat-
tlefield or disaster situations, where surgeons may be too precious to
put at risk on the front lines? And what if the demand for some specialized procedure is widely scattered around the world, but the necessary skills are only available in a couple of major centers? The need
to provide service under these sorts of conditions has motivated
intensive research into the possibility of telesurgery, and the development of some impressive prototype systems.[9]

So, yes, you can sometimes use telerobotics to be in touch—literally in *touch*—with distant service providers. But don't get too
excited about robotic limbs, tactile feedback devices, and shaking the
booty electric. Not yet, anyway.

■ The Teleservice Paradox

The limitations of telerobotics are instructive. Whatever the successes
of new teleservice systems, it remains true that *some* services—including many of the most humble—still depend on local presence of the

providers. Teleworkers need to get their dry cleaning done, and don't want to go too far for it. Electronic traders may operate globally, but the janitors who empty their wastebaskets and vacuum their floors have to be right there on the spot. The Telegarden notwithstanding, everyday gardeners still have to put their hands in the soil. Chefs have to get the food to the table while it's hot. Telehairdressing and teledentistry seem very far off in the future. Add it all up and you can quickly see that the goods and services produced within a city for local consumption—as distinct from what regional economists term the "export base"—are likely to remain a very significant percentage of the total.[10]

Thus concentrations of population and economic activity, once established, still have some powerful glue holding them together.[11] Digital delivery of educational, entertainment, medical, retail, financial, and other services can be expected to produce new patterns of service availability within and among these concentrations, but it will certainly not dissolve them. Indeed, something of a paradox emerges; hot spots of electronically mediated activity—such as Manhattan's Financial District, the City of London, or the teleworking-millionaire enclave of Aspen—become magnets for the low-wage service workers who do the sorts of things that computers and electronically controlled machinery cannot. And, of course, these concentrations of service workers become part of the attraction of such localities for the more privileged. It is a dirty little semisecret, then, that all of these high-flying places have large, far less interesting and attractive, low-rent counterparts somewhere nearby.

On this one, though, the fat lady is still clearing her vocal chords. Paul Krugman has suggested—and he is probably right—that the wrong side of the tracks will eventually have its revenge.[12] As networks spread, smart places proliferate, and software becomes more and more capable, the prices of information-related services will be driven down. At the same time, the value of manually performed services that cannot readily be automated or remotely delivered will correspondingly rise. Cooks, gardeners, nannies, and plumbers will do increasingly well.

Meanwhile, networking will radically change the way that small service providers operate. Taxi systems—with their individual operators coordinated by broking and telecommunications centers—have long shown the way. In the age of digital telecommunications, online service broking centers will provide directories, price and availability information, and recommendations. Instead of calling a plumber and getting put on hold or entangled in the phone tree, you may send out a software agent to find pipe wranglers with the necessary skills, check prices, availability, and references, and automatically set up an appointment. Instead of buying furniture at a local antique store, you will surf into a nationwide online auction site.

In sum, the spatial forces set in motion by teleservice are complex, and sometimes tug in different directions at once. They can produce both decentralizing and recentralizing tendencies. They can break the bond between local demand and local supply of services, but they can also reinforce the dominance of established service centers.

Electronic Fronts, Architectural Backs

Architecturally, the most striking consequence of teleservice is transformation of the traditional relationship between facade and back room. Many organizations are beginning to acquire electronic fronts and architectural backs.

Think, for example, of a retail store on an old-fashioned shopping street. The storefront presents the enterprise to the public, and the floor space immediately behind is where customers browse the stock, interact with sales staff, and make purchases. Behind that again is backroom stock and administration space which is *not* open to the public. Still further in the background, perhaps, is remote warehouse and head office space. Overall, there is a very clear hierarchy of visibility and public presence.

In the electronic equivalent of this store, though, the online interface takes over the functions of the street facade, the signage and display windows, and the retail floor; the software carries the full burden of mediating the organization's interactions with its customers. The backroom space remains; the need to store stock and accommodate administrative staff does not disappear. Locational constraints are loosened, however, and this backroom space can freely be distributed in whatever new patterns make practical sense. Furthermore, the buildings that provide this backroom space do not necessarily need to be in prominent, high-rent urban locations, nor do they need to perform representational roles. They can be remote and anonymous.

In an online bookstore, for example, the home page is the equivalent of the facade, and you find it by means of a search engine or by following pointers from other pages rather than come upon it by strolling along a street. The online catalog corresponds to the shelves of actual books, search engines and software agents help you to browse, and the online order form executes the functions of the checkout counter and cash register. Somewhere in the background, of course, there is a mega-warehouse, or a system of smaller warehouses distributed through the service area, where the books are physically stored, retrieved, packaged, and dispatched in the most traditional of fashions. And somewhere as well—maybe in a very different location, as determined by labor markets and telecommunications infrastructure— there are servers, call centers, and administrative offices.

Not surprisingly, the character and distribution of electronically administered backroom space varies according to the natures of the products and services that organizations provide. Highly perishable items that require very quick delivery, like hot food, need backroom space distributed throughout service areas; you cannot have a national pizza delivery center! Online supermarkets need warehouses located to support same-day delivery within metropolitan areas. Online bookstores, which rely on air and truck delivery, need large concentrations of backroom space at national and international transportation hubs. And financial services organizations, which do not deliver anything physically, can locate wherever rents and available labor forces make it attractive. Where backroom workers do not handle physical items, they can even be located in highly distributed telecommuter workspaces that have no spatial connection to customers whatsoever.

At the same time, an organization's electronic front changes the style and granularity of its public representation. Once, for example, banks were represented by branches located on Main Streets. Now, they are represented by much larger numbers of ATM machines, minibranches, and electronic home banking screens scattered in a very different, much more diffuse pattern.

■ Served and Serving Spaces Revisited

Most importantly, though, teleservice demands a new way of thinking about the organization of architectural space at both building and urban scales.

Back in the 1960s, Louis Kahn drew an influential distinction between the served and serving spaces of a building. Served spaces were the sites of important human activities, while serving spaces accommodated the support activities and equipment that the served spaces required. Thus a laboratory floor might be a served space, with adjacent plant rooms and exhaust shafts as serving spaces.

Since then, network technologists have learned to think in similar ways—and even to reinvent similar terminology. The Web, and comparable network structures, consist of client sites and server sites. Your home office might be a client site, for example, with your employer's intranet server supporting it.

Today, in the digital network era, the two traditions are beginning to converge. You might still relate smart served and serving spaces in the traditional way, by making them adjacent, but you might also establish their functional linkage through distant electronic connection. You can still partially read the functional organization of architectural space from floor plans and land use maps, but you must now look, as well, at the networking and the software.

In the electronically restructured cities of the twenty-first century, how will you choose between face-to-face and telecommunication? When will you want to travel to meetings, and when will you happily substitute remote connection? When will you communicate synchronously, and when will you decide to do so asynchronously? And how will our individual choices add up? What aggregate spatial and temporal patterns will emerge?

We will, I believe, plot our actions and allocate our resources within the framework of a new *economy of presence*.[1] In conducting our daily transactions, we will find ourselves constantly considering the benefits of the different grades of presence that are now available to us, and weighing these against the costs.

Elements of this economy of presence were in place and structured daily life in the cities of the past. But digital telecommunications infrastructure and smart spaces are now completing the system, and as a result they are introducing new possibilities and radically restructuring comparative benefits and costs.

■ The Cost of Being There

We don't usually think of it this way, but presence does consume resources and cost money. Typically, you pay more (in hotel charges or office rents, for example) to be present in places where many people would like to be than in places where few people want to be. And it costs you time and effort to get to places to meet people, conduct transactions, and see performances. Being in the right place, at the right time, can be expensive.

Before telecommunications technology began to change things, being "present" always meant having your body right there, in some specific location to establish the possibility of direct, face-to-face interaction. It entailed expenditure of resources for suitable real estate to get *together*, plus circulation or transportation systems to get you *there*. This, of course, was the essence of the ancient agora.

Under these conditions, propinquity—in both time and space—
was in great demand, and it became a very scarce and valuable resource.
There were prime times and prime locations. There were centers and
peripheries. Buildings and cities were carefully organized for efficient
space use and circulation.

▪ Traditional Limits

Furthermore, there were strict size limits; communities could not
become very large without beginning to fall apart. Members needed
to know one another, and to come together face to face to conduct
transactions and discuss matters of common interest, but the means to
these ends were finite. As both Plato and Aristotle pointed out in their
shrewd analyses of the functions and organization of cities, community life simply became impossible when there were too many people
trying to participate in it.[2] An agora could only be so big.

Beijing's vast Tiananmen Square vividly represents the functional
limits of traditional urban public space.[3] It is about one hundred acres
in extent, and if the crowds really pack themselves into it—as, on
occasion, they do—it can hold maybe a million people. This condition is not, however, conducive to multiway democratic discourse.
Tiananmen is mostly good for addressing the masses, and for
acclaiming leaders or putting bodies on the line to resist them.

▪ The Asynchronous Alternative

Even as the ancient Greek agoras were being built, though, a social
and cultural shakeup—one that would change things for ever—was
in the works. The first, primitive means for making visible marks on
surfaces had created the possibility of *externally* recording information; you saw something and drew it, or you heard something and
wrote it down. Senders and receivers of information no longer had

to be physically copresent to accomplish the transfer; separation in time was no longer an insurmountable barrier. A drawn or written message could be read long after the author had left the scene, and even—amazingly—after the author's death.[4]

Asynchronous communication had thus become possible. The long process of information disembodiment had begun. Economic, social, and cultural life could now be supported not only by movements and gatherings of people but also by the production, reproduction, storage, distribution, and subsequent use of inscribed information in human affairs. Connections and interactions among people, the communities that these exchanges created and sustained, and the forms of the cities that housed them, all started inexorably to change.

Lewis Mumford, for one, was convinced that this constituted the decisive urban moment. In his great work *The City in History*, he commented:

It is no accident that the emergence of the city as a self-contained unit, with all its historic organs fully differentiated and active, coincided with the development of the permanent record: with glyphs, ideograms, and script, with the first abstractions of number and verbal signs. By the time this happened, the amount of culture to be transmitted orally was beyond the capacity of a small group to achieve even in a long lifetime. It was no longer sufficient that the funded experience of the community should repose in the minds of the most aged members.[5]

In other words, cities came to depend on combining synchronous and asynchronous communication—speech and text, orator and scribe, live and Memorex, handshake and written contract, agora and archive. Each had its costs, advantages, and disadvantages, and these had to be weighed when there was a choice. It was the beginning of the economy of presence.

■ Information Mobilization

Asynchronous communication technologies evolved slowly at first, then with increasing pace as our own times approached. First, the storage media were heavy and difficult to transport, and were often integral parts of permanent structures; there were clay and stone tablets, and carved or painted markings on walls.[6] Religious and monumental buildings, in particular, were loaded with images and text, located centrally in communities, and designed to serve as foci of social, cultural, and spiritual life.[7] At this stage, readers usually came to the information instead of information to the readers.

But paper and similar lightweight media made inscribed information much more movable. First came the papyrus roll, then the more convenient codex book. Easel paintings, which could be bought and sold and shifted around, became increasingly popular alternatives to murals; as McLuhan once put it, they "deinstitutionalized" pictures.[8] Letters and bound manuscripts similarly mobilized text. Eventually, Aldus Manutius of Venice began to produce low-cost, portable, printed books.[9]

This new portability, allied with effective transportation, created the conditions necessary for introduction of public mail systems. These had their roots in the horseback courier systems that had been set up by kings and emperors from earliest times. Cyrus, Emperor of Persia, had such a system in the sixth century B.C. Similar arrangements served the Roman Empire and that of Charlemagne. From the sixteenth century, the mail systems of European monarchs got into the business of carrying letters for private citizens. By the nineteenth century, efficient and widely available public mail services were providing an increasingly indispensable way to communicate asynchronously—but relatively quickly—over considerable distances; messages could travel by stagecoach, steamship, rail, or even pony express.

As modern nation-states emerged, they established national mail systems as government monopolies or near-monopolies, and they entered into treaties for international exchange of mail. The resulting global network was the first of many such large-scale information distribution systems that were to follow. And, though it was many orders of magnitude slower than today's digital telecommunications systems, it possessed many of their essential structural features.

■ The Beginnings of Remote Interaction

Things had thus been reversed; information now came to the readers rather than readers to the information. As the novels of Austen, Dickens, and Trollope illustrate, the postman had begun to play an important role in sustaining social life. Businesses exchanged orders and invoices by mail. Educated professionals began to find that far-flung communities of interest—supported by correspondence—vied with local society for their attention and allegiance.[10] And the first of the lone-eagle telecommuters appeared; in the 1880s, Robert Louis Stevenson could settle on a remote Samoan island, continue to make a living as a prolific and successful author, and keep in contact with his many friends and acquaintances—all because the ships on the Sydney-to-San Francisco run stopped monthly at Apia to pick up and drop mail.

By the middle decades of the twentieth century, John Dewey could look back over the millennia and reflect:

It seemed almost self-evident to Plato—as to Rousseau later—that a genuine state could hardly be larger than the number of persons capable of personal acquaintance with one another. Our modern state-unity is due to the consequences of technology employed so as to facilitate the rapid and easy circulation of opinions and information, and so as to generate constant and intricate interaction far beyond the limits of face-to-face communities.... The elimination of distance, at the base of which are physical agencies, has called into being the new form of political association.[11]

Mobilization of information had thus added a new dimension to the economy of presence. Systems of social and economic integration could emerge at larger scales. And within them, there were choices to be made between travel to face-to-face meetings and reliance on remote communication.

■ Download a Life!

In all of this, the technical properties of media and messages proved to be crucial. *Durable* messages could transcend time, *compact* messages could minimize the storage space needed to accomplish this, and *lightweight* messages could conquer distance by reducing the difficulty and cost of transportation. Libraries and mail services could hardly have evolved to their present levels of sophistication and effectiveness if we were still relying on inscription of heavy, bulky stone tablets.

Paper prepared the way, but it was the nineteenth century's growing command of electromagnetism that finally cracked the problem of dematerializing messages and speeding them over long distances. It yielded the then-amazing possibility of encoding a signal at one end of a wire, transmitting it, and finally decoding it at the distant other end. This opened up the first era of electronic telecommunications—of the telegraph, the telephone, then (without even the wires) of radio and television broadcasting. In business and industry it enabled a revolution in coordination and control,[12] and culturally it spawned the early global village that McLuhan so vividly chronicled.

The next big advance—packet switching—was neither a new recording and storage technology nor a new transmission technology, but a means of effectively *managing* high-speed, high-volume information flows through telecommunications networks. It first appeared as an experimental technology in the 1960s, proliferated in the '70s and '80s, and had become indispensable by the '90s. In a couple of decades, it changed our whole way of thinking about telecommunications.[13] It gave us the ARPANET, Ethernets and other forms of local-area networks, the Internet, and the World Wide Web.

Unlike telephone and cable television networks that operate synchronously, packet-switched networks are primarily designed right from the beginning for asynchronous transfer of digital information.

The key idea is to break messages up into small "packets" of data, each one of which is tagged with information specifying its intended destination.[14] A packet may consist of several short messages, or a long message may require several packets.[15]

These addressed packets are directed through the network—typically via intermediate electronic devices, just as a mailed envelope might pass through several post offices—and they are eventually reassembled in correct sequence at the receiving end.[16] It is a lot like tearing out the individual numbered pages of a book, mailing them in separate envelopes to the same address, then putting them back together when they arrive—except that the disassembly and reassembly operations are automatic and invisible to the user.[17]

This idea might not have had such revolutionary potential if computers had remained scarce and expensive devices—as they were in the 1960s when it was first implemented. (Most of us would not care about the esoteric details of switching technology in specialized laboratory and business contexts.) But, in combination with silicon—inexpensive computer memory chips, processors, and optical fiber linkages—it was explosive. It opened up the possibility of today's immense networks in which vast quantities of digital information were stored in distributed fashion, in which information could rapidly be moved from any node to any other, and in which machine intelligence is used to manage and interpret inconceivably complex information flows.

■   Modes and Options

By this point, the economy of presence had been fully fleshed out; we now had the means to interact with one another both locally and remotely, both synchronously and asynchronously, and in all possible combinations of these.

Imagine, for example, that you want to get some information to a colleague. What are your options? They are schematically summarized in the following table.

	Synchronous	Asynchronous
Local	Talk face-to-face	Leave note on desk
Remote	Talk by telephone	Send email

First of all, you can simply stroll down to her office and discuss the matter face to face. This puts the two of you physically in the same place at the same time; in other words, it is local, synchronous communication. It is supported by the architectural arrangements—the provision of suitable space, desks, chairs, and conference tables. If her office is a smart place, you may augment your verbal interaction electronically—by making a videoprojected presentation from your laptop computer, for example.

If she is not there, you can leave a written note on her desk (or stick it on her computer screen), so that she can read it at some later point. This depends on the two of you getting to the same place, but you do not have to be there at the same time; thus it is a case of local asynchronous communication. It requires a convenient recording and storage technology, and the recipient must be able to find the message easily. In its more elaborate forms, it makes use of notice boards and bulletin boards, library shelves, and devices such as vending and ATM machines that allow the controlled asynchronous transfer of material artifacts.

A third alternative is to call her telephone extension. If she is there, and she answers, you interact remotely and synchronously. In this case, the supporting technology takes the form of a telecommunications system. Instead of a telephone, it could, of course, be a videoconferencing system or a shared virtual environment.

Finally, you might interact remotely and asynchronously by exchanging email or voicemail. This requires some combination of telecommunications and recording and storage technology—perhaps as simple as an answering machine attached to a telephone handset, or as elaborate as the Internet.

■ Costs and Benefits

How do you choose among these alternatives? It turns out that they differ considerably in their costs, advantages, and disadvantages, so you typically evaluate them according to the demands of particular contexts and situations.

Face-to-face provides the most intense, high-quality, potentially enjoyable interaction. It is not constrained by storage capacity, telecommunications bandwidth, or interface limitations. But it is also by far the most expensive option, both in direct cost and opportunity cost; it requires travel, and it consumes real estate, often in expensive, central locations. Most importantly, it consumes your attention; you only have a limited amount of time available in your day for meeting with people, and it demands some of this. So it makes sense in contexts where the importance of the interaction justifies the high cost.

Asynchronous communication is far less direct and intense, and it filters out a lot; reading Oscar Wilde is certainly not the same as meeting Oscar Wilde. However, it opens up the possibility of communicating across gaps in time, reduces unwelcome interruptions, and makes life easier by removing the need to coordinate schedules and allowing you to conduct your end of the interaction whenever you want. Opportunity costs are effectively reduced because you do not have so many potential interactions competing for your attention at hours of peak activity. In many contexts, these advantages greatly outweigh the disadvantages; although we may miss the human interaction with a bank teller, most of us, most of the time, prefer the asynchronous convenience of the ATM.

Remote communication also subtracts something; talking to your lover on the telephone (or even on a teleconferencing system) does not compare with being there in person. But it has the huge advantage of eliminating travel time and costs. Thus we tend to prefer it in contexts where speed and low cost are crucial, and the loss of immediacy does not matter too much.

Remote asynchronous communication goes to the extreme of separating participants in *both* space and time. Thus an email message is far less personal than a face-to-face meeting or even a telephone call. But it may also be far more convenient and much less costly— particularly where distances and time zones are involved. Many busy people, today, can effectively handle dozens or even hundreds of email interactions in a workday—with correspondents scattered all around the world—but they could not deal with more than a small fraction of these through face-to-face meetings or the telephone.

The advantages, disadvantages, and costs of these various interaction modes may be summarized as follows.

	Synchronous	Asynchronous
Local	Requires transportation Requires coordination Intense, personal **Very high cost**	Requires transportation Eliminates coordination Displaces in time Reduces cost
Remote	Eliminates transportation Requires coordination Displaces in space Reduces cost	Eliminates transportation Eliminates coordination Displaces in time and space **Very low cost**

In preliterate societies, the action was all in the local-synchronous quadrant; there was no alternative, and the associated costs severely constrained the sizes and forms of settlements. With literacy, as Mumford and others have noted, a significant amount of human interaction shifted to the local-asynchronous quadrant, and cities began to develop into their characteristic modern forms. With telecommunications, the remote-synchronous quadrant opened up, the scales of organizations and social units grew, and the long process of globalization began in earnest.

And much more recently, with the development and large-scale deployment of digital networks, there has been a rapid, massive shift of activity across the diagonal of our table to the very-low-cost, remote-asynchronous quadrant. This has been the most fundamental effect of the digital revolution.

■ Making Choices

How far will it go? Will the convenience and low cost of networked, remote, asynchronous communication simply drive out all the alternatives?

On the evidence so far, that seems highly unlikely. Instead, all the different modes will have their appropriate roles, and we will distribute our choices across all four quadrants, according to our needs and our willingness to pay the associated costs in particular contexts. To illustrate this point, let us consider how you might choose among the alternative ways of getting a message to a colleague.

Partly, of course, it depends on the nature and importance of the topic. If it is of crucial importance, and you think that personal presence really matters, then you will make the effort to leave your office and you will be willing to use up some of the limited and precious time that you have available for meeting with people. If it is much less important, though, you will probably be content with one of the quicker, cheaper, and less direct modes, and you will conserve your time and energy for other, higher-priority purposes.

In the extreme case, when a matter is highly sensitive and confidential, you might not want to leave any records that could be discovered by others or send any messages that could be overheard or intercepted. Face-to-face communication, in a place safe from eavesdropping, thus becomes the best option. (This is why mob-frequented bars have back rooms, spies talk with the shower running, and high-level lawyers and business executives need fancy but noisy Manhattan restaurants.)

Your choice may also be influenced by the prior relationship that you have established with your colleague. If you have known and trusted each other for a long time, then a brief email message may suffice even when the topic is a very sensitive one, since you can be confident that your written words will not be misinterpreted. If you are not very well acquainted, though, you may feel a greater need to minimize the danger of misunderstanding or bruised feelings by meeting face to face.

What if your colleague has a virulent case of the flu, has an office that smells of rancid fast food, ancient sneakers, and stale cigarette smoke, or is likely to take violent exception to what you have to say? Since the telephone is less risky and unpleasant than face-to-face under these circumstances, you may well prefer to chicken out and use it. If you want to avoid any sort of confrontation, then sending email is even better. (As Paul Simon might want to note, it's yet another way to leave your lover.) Alternatively, you may feel that it is cowardly and irresponsible to do this, and decide that it is a better course to go and face the music.

Your location at the time, and the condition that you happen to be in, matter as well. If it is a short walk to the other office, then the extra effort required to meet face to face is very small, and you may feel that it's well worth it—even for a casual discussion of a topic of minor importance. If the other office is distant, though, the cost for the same benefit is higher, and you may be more inclined to use the telephone or email. If you are young and healthy, then a walk to the other office may be easy and enjoyable, but if you are aged and infirm, or if you have a broken leg, then walking becomes a greater effort, and you need a greater benefit to justify it. If you both work the same hours, then synchronous communication may be easy to accomplish, but if you work different shifts, or if one of you happens to be on the road and in a different time zone, then the convenience of communicating asynchronously through email, voicemail, or fax may outweigh the loss of immediacy that is inherent in these means.

Then, there is the question of what else you have to do. When there are conflicting demands for your presence, you cannot resolve these by physically being in two places at once, but you can often divide your presence electronically. If you have to be at home to take care of a sick child, for example, you can still communicate with your colleague by telephone and electronic mail. (This division is made possible, in part, by the remarkable human capacity to process different streams of information in parallel; you can watch your child while you listen to someone on the telephone. It also exploits the fact that switching electronic connections among locations is much faster than physically traveling back and forth among widely scattered sites.) If there are few simultaneous demands for your presence, then you may be able to satisfy most of them by actually being there on the spot.

Conversely, if you have to try to satisfy many simultaneous demands, you will be forced to rely much more heavily on remote and asynchronous communication; that is why very busy CEOs often depend so heavily on email.

You may be concerned about the indirect, unspoken purposes of the interaction as well as the ostensible ones. If you are the boss, for example, then you can emphasize the importance of a message, or signal sympathy or support, by paying a personal visit to the office of a junior subordinate rather than telephoning or sending email. At the same time, you probably learn little about the subordinate through an email exchange, more through a telephone discussion, and quite a lot through an intense face-to-face discussion.

And you may simply be concerned about maintaining an appropriate balance in your life. If you have been spending too much time telephoning and emailing, the lack of direct human contact may be making you bored and lonely.[18] In that case, it is better to get out of your office and take a walk down the corridor.

Finally, you may realize that the different modes of communication available to you are not just discrete alternatives but can sometimes be used very effectively in combination. So you might pick up the telephone to arrange a face-to-face meeting. Or you might access your colleague's online calendar to find the times when she is available for a telephone call or a meeting. You might even instruct a software agent to negotiate with her software agent to find a mutually convenient time for you to get together. Sometimes these sorts of combinations can produce the "pen pal" phenomenon; you might initiate a contact through email, progress to telephone conversations, and eventually decide that it would be worthwhile to meet face to face.

■ The Persistent Power of Place

The character and quality of your colleague's office may matter as well. If it is a pleasant place to be, and if it offers the atmosphere and privacy that you need for conducting your business, then the chances are greater that you will go there. But if it is a crowded, squalid cubby-hole, you may feel just as well served by telephoning or sending email.

Since place retains this sort of power, it follows that place-based enterprises will compete for our presence, attention, and dollars in a digitally mediated world by attempting to add as much value as possible to the face-to-face experiences that they offer. They will emphasize the unusual, the elsewhere unobtainable, and the things that cannot (at least yet) be pumped through a wire.

Movie theaters, for example, will offer bigger screens, better sound systems, and more intense engagement with fellow viewers than is possible through home video-on-demand. Bookstores threatened by Internet sites will fight back by creating a welcoming ambiance for book lovers, by providing cappuccinos and cozy places to hang out, and by emphasizing the sensual pleasures of snapping the spines and riffling the pages of beautifully produced volumes. And old-line clothing retailers will promote the advantages of directly touching and trying on the merchandise.[19]

Fast food outlets may embrace online ordering and home delivery, but upmarket restaurants will continue to go for distinctive, place-based experiences. Maybe you could take Spago out of Hollywood or the Hollywood out of Spago, but that would defeat the whole purpose of the place. And you really do have to *be* there to get what it uniquely has to offer.

Local food shops that want to compete with online supermarkets will pitch themselves to the senses through foodie-magnet displays of produce, nostril-tickling aromas of coffee, spices, and baked goods, and tempting tasting stations at every aisle. Those very same shoppers who save time during the week by ordering their detergent and toothpaste from an online supermarket may allocate some of their weekend leisure time to visiting a sophisticated wine and cheese store.

Traditional kinds of public spaces will continue to flourish where they have out-of-the-ordinary, difficult-to-replicate local attractions to offer. Online commerce may diminish the capacities of downtown shopping streets to attract the public, for example, but it's still hard to beat the beach on a warm Sunday—and teletransactions may free up more time to go there. Electronic delivery will allow you to hear practically anything, anywhere, anytime you want it, but that will not diminish the thrill of having your eardrums assaulted by the mega-amplified Rolling Stones as you gulp Rolling Rock in a wildly reverberating football stadium somewhere. The opera at La Scala is pretty good, too.

■ No Substitutions

As this simple thought experiment suggests, then, the various forms of local presence and telepresence, and of synchronous and asynchronous communication, have similar and sometimes overlapping uses but are not exact functional equivalents. They add value to interactions and transactions in different ways, consume resources of different kinds and at different rates, and are feasible under different sets of conditions.

So they do not straightforwardly substitute for each other, and we should not expect a wholesale replacement of face-to-face interaction by electronic telecommunication, as technoromantics sometimes suggest and traditionalists often fear. Instead, we are likely to discover that different people, in different contexts, responding to different demands, subject to different constraints and with different resources at their disposal, will choose to conduct their interactions in widely varying fashions. They will set their priorities, make their tradeoffs, and ultimately arrive at different balances of materiality and virtuality, and of telecommunication and transportation.

As a result, cities will evolve down varying paths. Global cities like New York and London will, no doubt, seek to strengthen their positions as command and control centers by investing in advanced telecommunications infrastructure and building smarter and smarter workplaces. Attractive residential locations—including resort and recreation centers—will become denser with live/work dwellings and teleworkers. Communities that have been marginalized through isolation or poverty will try to improve their conditions through remote education, telemedicine, and other kinds of electronically delivered, low-cost services. Advanced technopoles with high labor costs, such as Silicon Valley, will be avid buyers on the electronically mediated global labor market; cities with pools of lower-cost but highly skilled labor—the Delhis, Bangalores, and Kingstons of the world—will be sellers. Cities with transportation and package delivery hubs will end up playing key roles in new electronic commerce systems. Centers of culture, entertainment, research, and scholarship will become more specialized; they will focus on what they do uniquely well, while electronically importing whatever other intellectual resources they may need. All will seek the advantages that make the most local sense.

It is a mistake to overgeneralize, as futurist gurus have been prone to do. The diverse architectural and urban forms of the future will surely reflect the balances and combinations of interaction modes that turn out to work best for particular people, at particular times and places, facing their own specific circumstances within the new economy of presence.

In the now-fading industrial era, we have made heavier and heavier demands upon our cities. As a result, they have grown ever larger, more crowded, more stressed and strained, and more desperately choked with traffic and pollution. The much-quoted *Agenda 21* statement anticipates that, by the year 2025, the world's cities will accommodate 60 percent of its population.[1] It is frighteningly obvious that we cannot continue down this path for very much longer.

But the digital revolution, together with the new economy of presence that is emerging from it, offer us some hopeful alternatives. Virtuality now vies with materiality. Travel is no longer the only way to go. And human intelligence is augmented, on a vast scale, by the silicon/software partnership. As a result, familiar urban patterns have lost their inevitability.

■ Five Points

In their place, we can create e-topias—lean, green cities that work smarter, not harder. Their basic design principles may be boiled down to five points—oversimplified, no doubt, but useful to hold in the mind. They are:

1. *Dematerialization*
2. *Demobilization*
3. *Mass customization*
4. *Intelligent operation*
5. *Soft transformation.*

By following these principles we can potentially meet our own needs without compromising the ability of future generations to meet theirs.[2] We can apply them at the scales of product design, architecture, urban design and planning, and regional, national, and global strategy.

Here's how.

■ Dematerialization

When a virtual facility like an electronic home banking system substitutes for a physical one like a branch bank, there is a net dematerialization effect; we no longer need so much physical construction, and we no longer have to heat and cool it. Replacement of big, physical things by miniaturized equivalents—as when silicon chips begin to do the job of vacuum tubes, and hair-thin fiber optics substitutes for heavy copper cables—accomplishes much the same result. And there are analogous benefits when we separate information from its traditional material substrates; an email message, read on the screen, does not consume paper.

Furthermore, we can win coming and going. If we never produce a material artifact, and make use of a dematerialized equivalent instead, it never turns into waste that has to be managed. A used bit is not a pollutant!

All this is becoming so obvious that the term "weightless economy" has gained increasing currency among economists and business commentators.[3] (Before long, of course, "weightless" will seem as quaintly anachronistic as "horseless," "wireless," and "zipless.") And we can no longer take the architectural implications lightly. Now, less really can be more.

Until recently, so-called green architecture has typically been pursued under the assumption that physical construction is unavoidable and the task is therefore to carry it out it as efficiently as possible. Consequently, it has rarely amounted to much more than well-intentioned tinkering with building massing and orientation, material choices, and energy systems, and it has not had the large-scale impacts that its proponents have sought. Today, though, the new economy of presence affords us the possibility of repeatedly asking the more radical questions, "Is this building really necessary? Can we wholly or partially substitute electronic systems instead?"

The overall effect of electronic dematerialization does depend, to be sure, on the levels of resource consumption required in the manufacture and operation of computational devices. These are not insignificant.[4] Semiconductor manufacture consumes energy, photochemicals, acids, hydrocarbon-based solvents, and other materials. IBM estimated

that junked computers were taking up a couple of million tons of
U.S. landfill at the turn of the century. It was also estimated that computers were consuming ten percent of the total U.S. electric power
supply. But these levels are certainly modest enough to promise very
substantial savings of resources through substitution of electronics for
construction. And the trend is toward smaller devices, greener manufacture, and lower power consumption.

■ Demobilization

We also conserve resources whenever we wholly or partially substitute telecommunication for travel. In general, moving bits is immea-
surably more efficient than moving people and goods. The savings
show up in reduced fuel consumption levels, lower pollution levels,
lessened need for occupation of land by transportation infrastructure,
cutbacks in vehicle manufacture and maintenance expenditures, and
shortening of time spent in traveling.

Interest in conserving resources and reducing pollution through
demobilization first emerged during the OPEC oil crises of the
1970s, when it was widely expected that telecommuting within the
framework of existing urban patterns might yield significant savings.
It soon became evident, however, that telecommunication could not
serve as a surrogate for transportation in such a straightforward way.[5]
The interactions of people, bits, and atoms turn out, as we have seen,
to be far too complex and subtle for that.

Despite this initial disappointment—in retrospect, the dashing of
naive early hopes—the new economy of presence *does* open up the
possibility of significant resource conservation through demobilization. In part, this is a matter of incentives; as Peter Hall has observed,
"If governments respond by raising the real cost of driving, either
overall or at peak times (through road pricing), or by restraining traffic by restricting the amount of space for driving or parking, then
(other things remaining equal) there will be a search for substitutes
for personal transport, at least for a certain proportion of journeys.

We might foresee some routine workers, especially part-time workers, working entirely from home or neighborhood workstations, while other workers practiced flexitime, coming to centralized meeting-places for some hours or days each week; thus reducing the overall volume of traffic, and also redistributing it away from the congested peaks."[6] The real key, though, is not to look for simple, direct substitutions, but to take advantage of telecommunication to create new, finer-grained, inherently more efficient urban patterns.

Specifically, the live/work neighborhood promises to reduce the wasteful daily commutes that have resulted from the typical industrial-era separation of homes and workplaces. Trips to nearby neighborhood facilities can be on foot or by bicycle. And electronic distribution of services eliminates longer trips to intermediate access points; you can download a movie from a national server, for example, instead of driving to the video store at the regional shopping center.

One promising strategy, then, is to pursue the development of poly-centric cities composed of compact, multifunctional, pedestrian-scale neighborhoods interconnected by efficient transportation and tele-communication links.[7] These units might be arranged linearly, along public transport spines.[8] By remixing homes, workplaces, and service facilities in this way, we can seek a more sustainable balance of pedes-trian movement, mechanized transportation, and telecommunication.

■   Mass Customization

Dematerialization and demobilization are the most obvious conserva-tion strategies within the new economy of presence, but they are not the only ones. We can also pursue the more subtle benefits of mass customization.[9]

The dumb machines of the industrial era gave us economies of standardization, repetition, and mass production, but the smart machines of the computer era can now provide us with the very dif-ferent economies of intelligent adaptation and automated personal-ization. We can employ silicon and software on a vast scale to enable automatic custom delivery of just what is required in particular contexts, and no more.

On any given morning, for example, you are very unlikely to read all the pages of your newspaper; most of them are simply wasted on you—unless you have a new puppy, or need to line bird cages. But an electronically delivered, home-printed, personalized newspaper system may have a profile of your interests and use it to select and print out just those articles and classified advertisements that you are likely to want to see. This strategy gobbles fewer trees to begin with, and it produces less waste in the end. In principle, it could be implemented by applying a human labor force to the task; in practice, there aren't enough editors and layout artists, and they could not work fast enough anyway. It depends upon the availability of inexpensive computation and telecommunication.

Similarly, your car just sits in garages and parking lots most of the time, and ties up resources to no useful effect. By contrast, a sophisticated, electronically managed rental and distribution service might provide just the type of vehicle you wanted—sometimes a minivan and sometimes a sporty two-seater—wherever and whenever you needed it. There may be more to gain from cleverer management of vehicle fleets than from trying to build ever-more-efficient privately owned automobiles.

We can get analogous benefits from intelligent, electronically mediated management of other transportation resources. When taxis are equipped with position-sensing devices, the nearest one can automatically be sent to answer a call. When transportation companies are electronically interconnected to one another and to their clients, they can efficiently coordinate pickups, improve load factors and back-haul planning, and reduce warehousing requirements through just-in-time delivery.[10] When intelligent vehicles run on smart road networks, routes can be optimized to minimize travel time and reduce congestion.

Old-fashioned mass production and electronically mediated mass customization turn out to have vividly contrasting formal implications. At the height of the industrial era, in the 1920s, Henry Ford rigorously standardized the Model T and famously offered it in any color —as long as it was black. Similarly, Mies van der Rohe standardized building modules, construction elements, and details, explored the spare poetry of simple shapes and regular repetition, and produced steel-and-glass buildings that were—well—black. Other heroic modernists preferred white but were equally entranced with the dumb-machine logic of standardization and repetition. But there was a nagging contradiction; one size never really fitted all. If you made a structural frame from uniform elements, some would be wastefully overdesigned. If you standardized a building's fenestration, some windows would appropriately mediate the varying interior and exterior conditions, but others, inevitably, would not.

Today, though, information-era projects such as Frank Gehry's Guggenheim Museum in Bilbao have begun to demonstrate a radical new resolution of the problem; they exploit the capabilities of computer-controlled production machinery to create compositions of nonstandardized, nonrepeating elements that respond precisely to their particular functions and contexts. The complex results are far from arbitrary and irrational, as unregenerate old Miesians like to complain, but responsive to a more subtle and sophisticated rationality. And, of course, they jolt our sensibilities by generating an astonishing new kind of spatial and material poetry.

At long last we can get it right. Thanks to the availability of inexpensive machine intelligence and ubiquitous telecommunications, we no longer have to choose continually between the unappealing alternatives of either standardizing and wasting resources or customizing and making production impossibly difficult.

Intelligent Operation

Much the same logic applies to those consumable resources that flow through pipes and wires—water, fuel, and electric power. By putting more intelligence into devices and systems that require these resources, we can minimize waste and can introduce dynamic pricing strategies that effectively manage demand and encourage thriftiness.

A really dumb, low-tech irrigation system, for example, relies on human gardeners to turn on the faucet and point the hose in the right direction. A simple automated system may be driven by a clock, so that it sprays water at regular intervals—even when rain is falling. A smarter system may be controlled by sensors, so that it dispenses water only when conditions indicate that supplementary moisture is necessary. But a *really* smart system should monitor both its environment and water availability levels, learn to predict irrigation needs, and automatically satisfy these needs without wasting water or making heavy demands when the supply is restricted.

Similarly, an elementary electrical system allows the lights and appliances in a house to be switched on and off. Slightly more sophisticated systems put some of the switches on timers, so that you don't have to be around to operate them and you don't waste electricity when the place is empty. With the addition of simple sensors, you can create a system that conserves energy by switching off the lights in rooms that are unoccupied for a while. (Unfortunately, they may also do it when you are just quietly sitting and thinking.) For maximum efficiency, though, you need a system that learns how you live, discovers patterns of dynamically varying electricity pricing, and optimally operates your lighting, heating and air conditioning, and appliances according to the predictive model that it maintains and continually updates.

This sort of automation is not about "labor saving"—the sales slogan for early domestic appliances. Nor is it motivated by infantile fantasies of being served hand and foot by infinitely compliant machines. Its goal is to create highly efficient, responsive markets for those scarce, consumable resources on which all human settlements depend. We have better things to do than trade in these markets, so we should leave it to our smart silicon surrogates—which will do better at it anyway.

■ Soft Transformation

In the hot spots of new development that emerge as the twenty-first century unfolds, there will undoubtedly be opportunities to create neighborhoods, and even whole new cities, that are organized to take

advantage of emerging opportunities for dematerialization, demobilization, mass customization, and intelligent operation. In most developed areas, though, the primary task will be one of adapting existing
building stock, public spaces, and transportation infrastructure to
meet requirements that are very different from those that guided
their initial production. These legacies of the industrial era, and of
even earlier times, will require transformation in order to function
effectively in the future.

Cities have experienced such transformations before. In particular,
the industrial revolution demanded provision of extensive industrial
areas, worker housing, downtown offices, and high-capacity transportation systems. Cities that could respond grew and prospered, while
many that could not went into decline. But the results of industrially
fueled growth and transformation were often, of course, extremely
destructive; old quarters were obliterated, architectural patrimony was
lost, railways and highways brutally divided urban tissues, and the
urban poor ended up living under miserable conditions. The transi-
tion costs were enormous.

Fortunately, the coming changes need not have such devastating
effects. Whereas new transportation infrastructure takes up large
amounts of space, frequently destroys areas of natural and historic
value, and increases noise and pollution, new telecommunications
infrastructure is far gentler and less obtrusive in its physical effects. It
will not need a Robert Moses; it can often be inserted almost invisi-
bly. In the beautiful old Italian city of Siena, for example, television
cabling was run throughout the historic quarter so that unsightly
aerials would not protrude above the rooftops; it now provides a
superb infrastructure for high-speed digital telecommunications.

Furthermore, as we have seen, electronically serviced space for
information work does not have to be concentrated in large contigu-
ous chunks, like the commercial and industrial zones of today's cities,
but can effectively be distributed through finer-grained urban fabric.
And unlike industrial facilities, it does not adversely affect the quality
of surrounding areas. In particular, it lends itself to accommodation
within the small-scale, endlessly varied spaces that characterize the
historic areas of older cities. This opens up promising opportunities

to go beyond nostalgic, rearguard preservationism; instead, we can reconnect, repurpose, and reboot valued but functionally obsolete urban fabric.

The path from what we have now to what we need in the future need not be one of cataclysmic change; we can follow the road of subtle, incremental, nondestructive transformation.

■   Our Town Tomorrow

In the twenty-first century, then, we can ground the condition of civilized urbanity less upon the accumulation of things and more upon the flow of information, less upon geographic centrality and more upon electronic connectivity, less upon expanding consumption of scarce resources and more upon intelligent management. Increasingly, we will discover that we can adapt existing places to new needs by rewiring hardware, replacing software, and reorganizing network connections rather than demolishing physical structures and building new ones.

But the power of place will still prevail. As traditional locational imperatives weaken, we will gravitate to settings that offer particular cultural, scenic, and climatic attractions—those unique qualities that cannot be pumped through a wire—together with those face-to-face interactions we care most about.

Physical settings and virtual venues will function interdependently, and will mostly complement each other within transformed patterns of urban life rather than substitute within existing ones. Sometimes we will use networks to avoid going places. But sometimes, still, we will go places to network.

NOTES

1   Marshall McLuhan, "The Alchemy of Social Change," Item 14 of *Verbi-Voco-Visual Explorations* (New York: Something Else Press, 1967). He went on to reinforce the point: "Any highway eatery with its TV set, newspaper, and magazine is as cosmopolitan as New York or Paris . . . . The metropolis is OBSO-LETE." He has not been alone in this view. For example, the eminent French architectural and urban theorist Françoise Choay (preface to *The Rule and the Model: On the Theory of Architecture and Urbanism* [Cambridge: MIT Press, 1997]) has suggested that the term "city" no longer correctly applies to our current urban environments, and that its use should be reserved for certain environments of the past. Choay is a lover of urbanity, and views this development with resignation and regret. But others proclaim good riddance; the conservative ideologue and techno-cheerleader George Gilder (*Forbes ASAP*, February 27, 1995, p. 56) has argued that "we are headed for the death of cities," which are nothing more, anyway, than "leftover baggage from the industrial era."

2   Among the classic statements of the traditional view as it had crystallized by the end of the 1950s are Lewis Mumford, *The City in History: Its Origins, Its Transformations, and Its Prospects* (New York: Harcourt Brace, 1961), and Jane Jacobs, *The Death and Life of Great American Cities* (New York: Vintage, 1961). Mumford and Jacobs represented opposing sides of a contemporary debate, and certainly would not have welcomed being lumped together, but from the perspective developed here they have far more similarities than differences. Changing conceptions of the city in the recent past are brilliantly surveyed in Peter Hall, *Cities of Tomorrow: An Intellectual History of Urban Planning and Design in the Twentieth Century* (Cambridge, Mass.: Blackwell, 1988). And Hall's *Cities in Civilization* (New York: Pantheon, 1998) is a sophisticated, up-to-date reworking of Mumford's themes.

1   Traditional Marxists, McLuhanites, and Silicon Valley futurologists have all tended to forms of technological determinism. In *Television: Technology and Cultural Form* (New York: Schocken, 1975), Raymond Williams fired an influential critical broadside against it, and powerfully influenced succeeding generations of social scientists—particularly on the left. For a more recent critique of the view of technology as agent, see Leo Marx, "Technology: The Emergence of a Hazardous Concept," *Social Research*, Fall 1997.

2   Technology is understood, here, in the sense put forward by Herbert
Marcuse in his famous 1941 essay "Some Social Implications of Modern Technology," reprinted in Herbert Marcuse, *Technology, War and Fascism: Collected
Papers of Herbert Marcuse*, vol. 1, ed. Douglas Kellner (London: Routledge, 1998),
pp. 39–65. Marcuse took technology to be "a social process in which technics
proper (that is, the technical apparatus of industry, transportation, communication) is but a partial factor. . . . Technology, as a mode of production, as the totali-
ty of instruments, devices and contrivances which characterize the machine age
is thus at the same time a mode of organizing and perpetuating (or changing)
social relationships, a manifestation of prevalent thought and behavior patterns,
an instrument for control and domination."

3   The phrase is from the first issue of *Wired* magazine, in 1993. To date, the
most detailed and comprehensive analysis of the economic, social, and political
dynamics of the digital revolution is Manuel Castells's masterly *The Rise of the
Network Society* (Oxford: Blackwell, 1996). Its technological underpinnings are
most clearly described in a trio of popular insider accounts from the mid-1990s:
Nicholas Negroponte, *Being Digital* (New York: Knopf, 1995); Bill Gates, *The
Road Ahead* (New York: Viking, 1995); Michael Dertouzos, *What Will Be* (New
York: HarperEdge, 1997). My own *City of Bits* (Cambridge: MIT Press, 1995)
suggested that architects and urban planners should sit up and take notice. For a
bleakly dystopic view, much at odds with all of these, see Paul Virilio, *Open Sky*
(London: Verso, 1997). And for a detailed analysis of potential downsides, see
Gene I. Rochlin, *Trapped in the Net: The Unanticipated Consequences of Computeri-
zation* (Princeton: Princeton University Press, 1997).

4   Our own age is not the first to experience the effects of such a combination. In *Novum Organum*, Francis Bacon famously noted that the inventions of
the compass (to get you there), gunpowder (to impose your dominance), and the
printing press (to tell all about it) had given the moderns a great advantage over
the ancients.

5   This point has become a truism among telecommunications analysts, and
there are numerous published accounts of progress so far and scenarios for the
future. The story of the ARPANET and the Internet is told in Katie Hafner and
Mathew Lyon, *Where Wizards Stay Up Late: The Origins of the Internet* (New York:
Simon & Schuster, 1996). A more technical account is given in Peter H. Salus,
Casting the Net: From ARPANET to Internet and Beyond (Reading, Mass.: Addison-
Wesley, 1995). The emergence of digital television is chronicled in Joel Brinkley,
Defining Vision: The Battle for the Future of Television (New York: Harcourt Brace,
1997). The early days of the World Wide Web are described in Robert H. Reid,
Architects of the Web: 1,000 Days That Built the Future of Business (New York: John
Wiley, 1997). One attempt at a comprehensive overview is provided in Wilson
Dizard, Jr., *Meganet: How the Global Telecommunications Network Will Connect Every-*

one on Earth (Boulder: Westview Press, 1997). For a cogent forecast of where it's all going, see Gordon Bell and James N. Gray, "The Revolution Yet to Happen," chapter 1 of Peter J. Denning and Robert M. Metcalfe, *Beyond Calculation: The Next Fifty Years of Computing* (New York: Springer-Verlag, 1997), pp. 5–32.

6   And that's just what silicon clearly promises. Beyond that, more exotic possibilities such as quantum computing are looming into view. We are a long way from running out of new ideas for miniaturizing, cranking up clock speeds, and increasing parallelism.

7   For an exposition of this view, see Thomas L. Friedman, "The Internet Wars," *New York Times*, April 11, 1998, p. A27.

8   They were, at least, proximate causes of development. One can argue, of course—and political economists are often inclined to do so—that these were subsumed by larger patterns of social and political causality. On the general role of infrastructure in the modern city-building process, see Josef W. Konvitz, "The Infrastructure," in *The Urban Millennium: The City-Building Process from the Early Middle Ages to the Present* (Carbondale: Southern Illinois University Press, 1985), pp. 131–146. On historical patterns of infrastructure growth, replacement, and decline, see Arnulf Grubler, "Evolution of Infrastructures: Growth, Decline, and Technological Change," chapter 3 of *The Rise and Fall of Infrastructures* (Heidelberg: Physica-Verlag, 1990).

9   For an argument along these lines—focusing specifically on the case of Palo Alto, California—see John Markoff, "Old Man Bandwidth: Will Commerce Flourish Where Rivers of Wire Converge?" *New York Times*, December 8, 1997, pp. D1, D13. And for some additional evidence from other contexts, see Andrew Gillespie and William Cornford, "Telecommunication Infrastructures and Regional Development," in William H. Dutton, ed., *Information and Communication Technologies: Visions and Realities* (New York: Oxford University Press, 1996), pp. 335–352.

10   Still, the bandwidth was very limited, so early telephone channels filtered out many of the nuances of speech—reducing voices to tinny caricatures. Hence the term "phonies" for imposters and con men who used this limitation, together with the masking of their faces and body language, to hide their insincerity.

11   See, for example, Agnès Huet and Jean Zeitoun, *Les téléports: Nouvelles places de marche sur les inforoutes* (Paris: L'Harmattan, 1995).

12   In addition to making available the necessary data communication facilities, the government has enhanced the competitiveness of software parks by developing wired, ready-to-occupy workspace for software businesses, streamlining regulatory processes, and providing tax incentives.

13   UNESCO, *World Communication Report: The Media and the Challenge of the New Technologies* (Paris: UNESCO Publishing, 1997), pp. 18, 70.

14   For an analysis of the early growth of the American system of cities, before telecommunications, see Allan R. Pred, *Urban Growth and the Circulation of Information: The United States System of Cities, 1790–1840* (Cambridge: Harvard University Press, 1973).

15   This system began with telegraph, telephone, and telex links. Then Reuters brought it into the computer network era with its Monitor service, launched in 1973; this provided minute-by-minute data on fluctuating exchange rates. By the 1990s, every trader's desk had a sophisticated computer workstation that allowed doing deals online, traders were toting currency-monitor pagers, and several firms—Reuters, Bloomberg, Dow Jones Markets, and Bridge—were locked in cutthroat competition for the financial information and trading system market.

16   For an introduction to the relevant technologies, see Annabel Z. Dodd, *The Essential Guide to Telecommunications* (Upper Saddle River, N.J.: Prentice Hall PTR, 1998).

17   Ferdinand Tönnies, *Community and Association* (London: Routledge & Kegan Paul, 1953; original 1887).

18   For a concise account of this development, see Joel A. Tarr, "The Evolution of the Urban Infrastructure in the Nineteenth and Twentieth Centuries," in Royce Hanson, ed., *Perspectives on Urban Infrastructure* (Washington, D.C.: National Academy Press, 1984), pp. 4–60. For a collection of useful case studies, see Joel A. Tarr and Gabriel Dupuy, eds., *Technology and the Rise of the Networked City in Europe and America* (Philadelphia: Temple University Press, 1988). The general role of networks in city building is discussed in Konvitz, *The Urban Millennium*.

19   Upbeat perspectives on these markets are offered by Frances Cairncross, *The Death of Distance: How the Communications Revolution Will Change Our Lives* (Boston: Harvard Business School Press, 1997), and John Hagel III and Arthur G. Armstrong, *Net Gain: Expanding Markets through Virtual Communities* (Boston: Harvard Business School Press, 1997). More jaundiced critics, such as Neil Postman, have suggested that new digital technology is, in fact, nothing more than a grand scheme to expand the market for the entertainment industry; see Neil Postman, "Future Schlock: 500 Channels and Nothing On," *Toronto Globe & Mail*, 1993.

20   On community networks, see Stephen Doheny-Farina, *The Wired Neighborhood* (New Haven: Yale University Press, 1996), and Douglas Schuler, *New Community Networks: Wired for Change* (Reading, Mass.: Addison-Wesley, 1996). On the Well, see Howard Rheingold, *The Virtual Community: Homesteading on the Electronic Frontier* (Reading, Mass.: Addison-Wesley, 1993). On Echo, see Stacy

Horn, *Cyberville: Clicks, Culture, and the Creation of an Online Town* (New York: Warner Books, 1998).

21   See for example Mike Jensen, "A Guide to Improving Internet Access in Africa with Wireless Technologies," International Development Research Council Study, August 31, 1996.

22   On the growing capabilities of satellite systems, see John Montgomery, "The Orbiting Internet: Fiber in the Sky," *Byte*, November 1997, cover story.

23   Karl Marx and Friedrich Engels, *The Communist Manifesto: A Modern Edition* (London: Verso, 1998), p. 40. In his introduction to this edition, Eric Hobsbawm points out that "idiocy" here refers not so much to "stupidity" as to something closer to the sense of the Greek *idiotes*—"narrow horizons" or "isolation from the wider society" (p. 11).

24   Nicholas Negroponte, "One-Room Rural Schools," *Wired* 6, no. 9 (September 1998), p. 212.

25   On the practical impact of early filtering technology, see Larry Guevara, "Plain or Filtered," *Educom Review* 33, no. 2 (March/April 1998), pp. 4–6.

26   As the Internet and the World Wide Web grew explosively in the 1990s, lawyers and legislators became increasingly aware of this, and started to try to sort out the issues that emerged as a result. See, for example, M. Ethan Katsh, *Law in a Digital World* (New York: Oxford University Press, 1995), and Brian Kahin and Charles Nesson, eds., *Borders in Cyberspace* (Cambridge: MIT Press, 1997).

27   The notion that "space" need not be understood in a strictly geometric sense, but can usefully be understood as a social production, was put forth by Henri Lefebvre in *The Production of Space*, trans. Donald Nicholson-Smith (Oxford: Blackwell, 1991; original 1974). Lefebvre's related writings on cities are collected in *Writings on Cities*, trans. and ed. Eleonore Kofman and Elizabeth Lebas (Oxford: Blackwell, 1996).

28   This was clear by 1997. In an article on the current financial troubles of *Wired* magazine, executive editor Kevin Kelly was quoted as remarking: "You can only be cool once ... I think we're going into a postcool period." And Bruce Sterling added: "In the early days of the digital revolution, it really was kind of a revolution, and therefore all things seemed possible. ... But, hey, after the revolution comes the provisional government. And quite commonly, the revolution eats its young, baby." See Amy Harmon, "Fast Times at Wired Hit a Speed Bump," *New York Times*, August 4, 1997, pp. D1, D8. By 1998 there was no doubt about it; *Wired* had been absorbed into the Conde-Nast magazine empire.

CHAPTER 2   TELEMATICS TAKES COMMAND

1   In *Terminal Architecture* (London: Reaktion Books, 1998), Martin Pawley has developed this point into an argument that buildings of the twenty-first century will have to be understood not as monuments but as terminals for information. Agreed. But he projects far bleaker consequences than I do.

2   William Gibson's *Neuromancer* (New York: Ace Books, 1984)—the novel that popularized the term "cyberspace"—is often taken simply as an evocation of electronically mediated disembodiment and placelessness. But it can more rewardingly be read as an allegory on the complex, reciprocal interrelationships of particular physical places (such as Chiba City) and virtual places, physical travel and electronic connection, and bodies and their electronic avatars.

3   These began to appear in the mid-1990s. Among the earliest were Microsoft's V-Chat, Intel's Moondo, Sony's Cyber Passage Bureau, IBM's Virtual World, Lycos's Point World, AlphaWorld, Worlds Chat, The Realm, and Utopia. For a useful survey, circa 1997–1998, see Bruce Damer, *Avatars! Exploring and Building Virtual Worlds on the Internet* (Berkeley: Peachpit Press, 1998). The term *avatar* comes from the Sanskrit, and has traditionally referred to the representation of Hindu deities by idols taking many different forms. I quote, for example, from the *Deccan Herald*, Thursday, August 27, 1998: "Musician Ganesha, cricketer Ganesha, armed Ganesha, dancing Ganesha, Afghan Ganesha, Chinese Ganesha, Japanese Ganesha, Samanvyaa Ganesha, decorated Ganesha, Ganesha as Shirdi Sai Baba, Ganesha in a 'Titanic,' sitting Ganesha, standing Ganesha, metal Ganesha, clay Ganesha, wooden Ganesha . . . in tune with the myriad names Lord Ganesha has been endowed with, idols of the deity of different varieties were worshipped in the City on Vinayaka Chaturthi on Tuesday."

4   Paul Saffo, "Sensors: The Next Wave of Innovation," *Communications of the ACM* 40, no. 2 (February 1997), pp. 93–97.

5   Krzysztof Wodiczko, *Critical Vehicles: Writings, Projects, Interviews* (Cambridge: MIT Press, 1999).

6   Bill Gates, "Plugged In at Home," in *The Road Ahead* (New York: Viking, 1995), pp. 205–226.

7   Robert Venturi, *Iconography and Electronics upon a Generic Architecture* (Cambridge: MIT Press, 1996).

8   H. Ishii, M. Kobayashi, and K. Arita, "Iterative Design of Seamless Collaboration Media," *Communications of the ACM* 37, no. 8 (August 1994), pp. 83–97.

9   P. Maes, T. Darrell, B. Blumberg, "The ALIVE System: Wireless, Full-Body Interaction with Autonomous Agents," *Communications of the ACM* 39 (Spring 1996).

10   J. Jacobson, B. Comiskey, et al., "The Last Book," *IBM Systems Journal* 36, no. 3 (1997). See also Neil Gershenfeld, "Bits and Books," in *When Things Start to Think* (New York: Henry Holt, 1999), pp. 13–25.

11   Many of these ideas have been experimentally implemented in the ambientROOM project at MIT's Media Laboratory. See Hiroshi Ishii, Craig Wisneski, Scott Brave, Andrew Dahley, Matt Gorbett, Brygg Ullmer, and Paul Yarin, "ambientROOM: Integrating Ambient Media with Architectural Space," and Andrew Dahley, Craig Wisneski, and Hiroshi Ishii, "Water Lamp and Pinwheels: Ambient Projection of Digital Information into Architectural Space," both in *Proceedings of CHI 98* (New York: Association for Computing Machinery, 1998).

12   John Underkoffler employs this approach in his Luminous Room project. See John Underkoffler, "A View from the Luminous Room," *Personal Technologies* 1, no. 2 (June 1997), pp. 49–59.

13   Pierre Wellner, "Interacting with Paper on the Digital Desk," *Communications of the ACM* 36, no. 7 (July 1993), pp. 87–96.

14   Hiroshi Ishii and Brygg Ullmer, "Tangible Bits: Towards Seamless Coupling of People, Bits, and Atoms," *Proceedings of CHI*, 1997, pp. 234–241. For further development of this idea, see John Underkoffler and Hiroshi Ishii, "Illuminating Light: An Optical Design Tool with a Luminous-Tangible Interface," *Proceedings of CHI*, 1998, pp. 542–549.

15   Myron Krueger, *Artificial Reality II* (Reading, MA: Addison-Wesley, 1991).

16   Osamu Morikawa and Takanori Maesako, "HyperMirror: Toward Pleasant-to-Use Video Communication System," *Proceedings of CSCW 98: ACM 1998 Conference on Computer Supported Collaborative Work* (New York: Association for Computing Machinery, 1998), pp. 149–158.

17   John Underkoffler, "The I/O Bulb and the Luminous Room," PhD dissertation, Media Arts and Sciences Program, MIT, 1998.

18   Ivan E. Sutherland, "A Head-Mounted Three-Dimensional Display," *Proceedings of the Fall Joint Computer Conference* (Washington, D.C.: Thompson Books, 1968).

19   For an extended critical discussion of the interrelationships among Alberti's rectangle, computer graphics, and virtual reality, see Jay David Bolter and Richard Grusin, *Remediation: Understanding New Media* (Cambridge: MIT Press, 1998).

20   C. Cruz-Neira, D. J. Sandin, and T. A. DeFanti, "Surround-Screen Projection-Based Virtual Reality: The Design and Implementation of the CAVE,"

Proceedings of SIGGRAPH 93 (New York: Association for Computing Machinery, 1993), pp. 135–142. See also T. A. DeFanti, D. J. Sandin, and C. Cruz-Neira, "A 'Room' with a 'View'," *IEEE Spectrum*, October 1993, pp. 30–33.

21   For a fast summary of research and prototype applications, circa 1996, see Larry Krumenaker, "Virtual Assembly," *MIT's Technology Review*, February/March 1997, pp. 18–19. For more detail see S. Feiner, B. MacIntyre, and D. Seligman, "Knowledge-Based Augmented Reality," *Communications of the ACM* 36, no. 7 (July 1993), pp. 53–62, and R. T. Azuma, "A Survey of Augmented Reality," *Presence* 6, no. 4 (1997), pp. 355–380.

CHAPTER 3   SOFTWARE: NEW GENIUS OF THE PLACE

1   GPS technology is not new, but miniaturization and price reductions have been making widespread, everyday use increasingly feasible. Receivers used to be bulky devices costing tens of thousands of dollars; by the late 1990s they had become handheld consumer items selling for a few hundred dollars.

2   For pioneering applications of this idea, see Andy Harter and Andy Hopper, "A Distributed Location System for the Active Office," *IEEE Network* 8, no. 1 (1994), pp. 62–70, and Roy Want, Bill N. Schilit, Norman I. Adams, Rich Gold, Karin Petersen, David Goldberg, John R. Ellis, and Mark Weiser, "An Overview of the ParcTab Ubiquitous Computing Experiment," *IEEE Personal Communications* 2, no. 6 (1995), pp. 28–43.

3   Joe Paradiso and Neil Gershenfeld, "Musical Applications of Electric Field Sensing," *Computer Music Journal* 21, no. 2 (1997).

4   See, for example, Alex P. Pentland, "Smart Rooms," *Scientific American*, April 1996, pp. 68–76. For further details of technical approaches see Michael Coen, ed., *Intelligent Environments: Papers from the 1998 AAAI Spring Symposium*, Technical Report SS-98-02 (Menlo Park: AAAI Press, 1998).

5   Neil Gershenfeld, *When Things Start to Think* (New York: Henry Holt, 1999), pp. 152–154.

6   Paul Saffo, "Sensors: The Next Wave of Innovation," *Communications of the ACM* 40, no. 2 (February 1997), pp. 93–97.

7   This estimate is given in Ted Lewis's "Binary Critic" column, *IEEE Computer*, September 1997.

8   For details of Bluetooth, see www.bluetooth.com.

9   For details of Jini, see www.sun.com/jini/ and www.jini.org. Other technologies that emerged around the same time, such as Motorola's Piano, Hewlett-

Packard's JetSend, and the HAVi specification for interoperability of home digital devices, deal with similar and related aspects of the interoperability problem.

10   For details of Java, see www.sun.com/java/.

11   Agent technologies and their applications are comprehensively surveyed in Michael N. Huhns and Munindar P. Singh, *Readings in Agents* (San Francisco: Morgan Kaufmann, 1998). A practical guide to agent development is provided in Michael Knapik and Jay Johnson, *Developing Intelligent Agents for Distributed Systems: Exploring Architecture, Technologies, and Applications* (New York: McGraw-Hill, 1998).

12   For a detailed exposition of the idea of specialized information appliances, and arguments in favor of them, see Donald A. Norman, *The Invisible Computer: Why Good Products Can Fail, the Personal Computer Is So Complex, and Information Appliances Are the Solution* (Cambridge: MIT Press, 1998).

CHAPTER 4   COMPUTERS FOR LIVING IN

1   Some of the more interesting possibilities are described in Steve Mann, "Wearable Computing: A First Step Toward Personal Imaging," *IEEE Computer*, February 1997, pp. 25–32. See also Thad Starner and Steve Mann, "Augmented Reality through Wearable Computing," *Presence* 6, no. 4 (1997). On the antecedents of wearables and bodynets, see Chris Hables Gray, ed., *The Cyborg Handbook* (New York: Routledge, 1995).

2   T. Zimmerman, "Personal Area Networks (PAN)," *IBM Systems Journal* 35 (1996), pp. 609–618. See also Neil Gershenfeld, "Wear Ware Where," in *When Things Start to Think* (New York: Henry Holt, 1999), pp. 45–61.

3   Donna J. Haraway, *Simians, Cyborgs and Women* (New York: Routledge, 1991). See also Gray, ed., *The Cyborg Handbook*, and N. Katherine Hayles, *How We Became Posthuman: Virtual Bodies in Cybernetics, Literature, and Informatics* (Chicago: University of Chicago Press, 1999).

4   Among the first were Xybernaut (which offered a head-mounted, voice-activated multimedia computer), ViA, and Teltronics.

5   Gordon Bell, "The Body Electric," *Communications of the ACM* 40, no. 2 (February 1997), pp. 31–32.

6   Mark Weiser, "The Computer for the 21st Century," *Scientific American* 265, no. 3 (1991), pp. 94–104. For more technical detail see Mark Weiser, "Some Computer Science Problems in Ubiquitous Computing," *Communications of the ACM* 36, no. 7 (July 1993).

7   George Fitzmaurice, "Situated Information Spaces and Spatially Aware Palmtop Computers," *Communications of the ACM* 36, no. 7 (July 1993).

8   See for example Scott Elrod, Gene Hall, Rick Costanza, Michael Dixon, and Jim Des Rivieres, "Responsive Office Environments," *Communications of the ACM* 36, no. 7 (July 1993), pp. 84–85.

9   See for example David Schneider, "Power to the People," *Scientific American* 276, no. 5 (May 1997), p. 44.

10   The idea of dynamic road pricing is to charge more for currently congested routes and less for uncongested ones. Singapore introduced such a system, based upon automated electronic monitoring of multi-lane roads, in 1998.

11   W. Wayt Gibbs, "World Wide Widgets," *Scientific American* 276, no. 5 (May 1997), p. 48.

12   On satisfying the potentially diverse environmental needs of multiple occupants, see Joseph F. McCarthy and Theodore D. Anagnost, "MusicFX: An Arbiter of Group Preferences for Computer Supported Collaborative Workouts," *Proceedings of CSCW 98: ACM 1998 Conference on Computer Supported Collaborative Work* (New York: Association for Computing Machinery, 1998), pp. 363–372.

13   See Michael C. Mozer, R. H. Dodier, M. Anderson, L. Vidmar, R. F. Cruickshank III, and D. Miller, "The Neural Network House: An Overview," in L. Niklasson and M. Boden, eds., *Current Trends in Connectionism* (Hillsdale, N.J.: Erlbaum, 1995), pp. 371–380, and Michael C. Mozer, "The Neural Network House: An Environment That Adapts to Its Inhabitants," in Michael Coen, ed., *Proceedings of the AAAI Spring Symposium on Intelligent Environments* (Menlo Park: AAAI Press, 1988), pp. 110–114.

14   This point is developed, with great verve and insight, in Reyner Banham, *The Architecture of the Well-Tempered Environment* (Chicago: University of Chicago Press, 1969).

15   The classic social dilemma of new, large-scale infrastructures is that they take time and money to build, so you cannot get them to everyone, everywhere, all at once. Do you privilege expediency and efficiency by building them incrementally, adding users as you go, and ignoring the short-term inequities that result? Do you insist on equity, and delay providing service to anyone until you can get it to everyone? Or do you look for some realistic compromise?

CHAPTER 5   HOMES AND NEIGHBORHOODS

1   This idea got a lot of airtime—particularly as a rather naively utopian dream of escape from the problems and dangers of the city—as the digital revo-

lution was gathering momentum in the 1980s. See, for example, Alvin Toffler, *The Third Wave* (New York: Bantam, 1980), and Joseph Deken, *The Electronic Cottage* (New York: Morrow, 1981).

2   In fact, as the digital revolution unfolded in the 1990s, demand for downtown office space was very strong in most major U.S. cities.

3   For an excellent, comprehensive overview of telework and telecommuting issues, see Jack M. Nilles, *Managing Telework: Strategies for Managing the Virtual Workforce* (New York: John Wiley, 1998). On European developments, see Mike Johnson, "EU Study on Teleworking," in *Teleworking. . . in Brief* (Oxford: Butterworth Heinemann, 1997), pp. 193–208. The research literature on telecommuting is now extensive. See S. L. Handy and P. L. Mokhtarian, "Forecasting Telecommuting—An Exploration of Methodologies and Research Needs," *Transportation* 23 (1996), pp. 163–190; P. L. Mokhtarian, "The State of Telecommuting," *ITS Review* 13, no. 4 (1990); P. L. Mokhtarian, "Telecommuting and Travel: State of the Practice, State of the Art," *Transportation* 18 (1991), pp. 319–342; P. L. Mokhtarian, "Telecommuting in the United States: Letting Our Fingers Do the Commuting," *TR News*, no. 158 (1992), pp. 2–7; J. M. Nilles, "Telecommuting and Urban Sprawl: Mitigator or Inciter?" *Transportation* 18 (1991), pp. 411–431; R. M. Pendyala, K. G. Goulias, and R. Kitamura, "Impact of Telecommuting on Spatial and Temporal Patterns of Household Travel," *Transportation* 18 (1991), pp. 383–409.

4   Historically, the internal organization of domestic space has reflected— among other things—differing resolutions of questions of centralization and decentralization. Does socializing take place in private living rooms or in public places? Are there private shrines in every house, or is worship a communal activity taking place at a central assembly point? Do you work at home, or do you go to a central workplace? Do you have a private garage, or do you park in a public structure somewhere nearby? For discussion of a wide variety of examples, see Amos Rapoport, *House Form and Culture* (Englewood Cliffs, N.J.: Prentice Hall, 1969).

5   The older *machiya* were beautiful wooden townhouses built on long, narrow blocks. Craftsmen displayed their wares on the streets in front of their homes. Today, the pattern continues with newer construction. Homes, shops, small factories, and restaurants are intricately interwoven; only a *noren* curtain hung in the doorway tells you that a particular house is open for business. This forms a particularly flexible urban fabric, which has served as the incubator for many of Kyoto's successful modern enterprises.

6   For an early warning, see P. Mattera, "Home Computer Sweatshops," *The Nation* 236, no. 13 (1983), pp. 390–392.

7   Distinctions among primary and secondary relationships were drawn by G. H. Cooley, *Social Organization* (New York: Scribner, 1909). They have, by now, become a Sociology 101 staple. They have usefully been elaborated and applied to electronically mediated situations by Craig Calhoun, "Computer Technology, Large-Scale Social Integration, and the Local Community," *Urban Affairs Quarterly* 22, no. 2 (December 1986), pp. 329–349, and "The Infrastructure of Modernity: Indirect Social Relationships, Information Technology, and Social Integration," in Hans Haferkamp and Neil J. Smelser, eds., *Social Change and Modernity* (Berkeley: University of California Press, 1992), pp. 205–236.

8   Melvin M. Webber, "The Post-City Age," *Daedalus* 97 (1968), pp. 1091–1110. See also R. F. Abler, "What Makes Cities Important," *Bell Telephone Magazine* 49, no. 2 (1970), pp. 10–15; and P. C. Goldmark, "Communication and Community," *Scientific American* 227 (1972), pp. 143–150.

9   A classic early formulation of this point is provided in T. C. Koopmans and M. Beckman, "Assignment Problems and the Location of Economic Activities," *Econometrica* 25, no. 1 (1957), pp. 53–76.

10   See for example Kerry Hannon, "A Long Way from the Rat Race: The Charms of Telluride Have Made a Telecommuting Town," *US News and World Report*, October 1995.

11   The transformation of former printing and copy shops into neighborhood business centers provides some evidence for a trend in this direction. See Laurie J. Flynn, "For the Officeless, a Place to Call Home," *New York Times*, Business Day, July 6, 1998, pp. D1, D4.

12   See, for example, Richard Sennett, *The Fall of Public Man* (New York: Knopf, 1976).

13   Robert Putnam, "Bowling Alone: America's Declining Social Capital," *Journal of Democracy* 6, no. 1 (1995), is only the latest in a long line of commentators to diagnose a loss of community in modern life and to locate the causes in some combination of urbanization, suburbanization, the automobile, and television.

14   Jane Jacobs, *The Death and Life of Great American Cities* (New York: Vintage Books, 1961). On New Urbanist prescriptions, see Peter Calthorpe, *The Next American Metropolis: Ecology, Community, and the American Dream* (Princeton: Princeton Architectural Press, 1993), Peter Katz and Vincent Scully, *The New Urbanism: Toward an Architecture of Community* (New York: McGraw-Hill, 1993), and David Mohney and Keller Easterling, eds., *Seaside: Making a Town in America* (Princeton: Princeton Architectural Press, 1991). For some very different proposals by Richard Rogers, see *Cities for a Small Planet* (Boulder: Westview Press, 1997).

NOTES TO PP. 80-82

15   On the emergence of Silicon Alley, and its discontents as well as its virtues, see Andrew Ross, "The Great Wired Way," *Any*, no. 22 (1998), pp. 57–61.

16   For an overview of this spatial pattern, see John Kain, "The Spatial Mismatch Hypothesis: Three Decades Later," *Housing Policy Debate* 3 (1993), pp. 371–460.

17   See Manuel Castells, "The Informational City Is a Dual City: Can It Be Reversed?" in Donald A. Schön, Bish Sanyal, and William J. Mitchell, eds., *High Technology and Low Income Communities* (Cambridge: MIT Press, 1998), pp. 25–42. On the effects of unequal access to information infrastructure, see William Wresch, *Disconnected: Haves and Have-Nots in the Information Age* (New Brunswick: Rutgers University Press, 1996). And on the general tendency to retreat into gated communities, see Edward J. Blakeley and Mary Gail Snyder, *Fortress America: Gated Communities in the United States* (Washington, D.C.: Brookings Institution Press, 1997).

18   Dual cities engendered by the simultaneous privileging and marginalizing effects of technological transformation have, in the past, been a favorite theme of novelists. Just think of Dickens, and his characteristic dramatization of the contrast by having protagonists move from one realm to the other. The digital revolution has spawned similar treatments by the cyberpunks. In his chillingly hilarious *Snow Crash* (New York: Bantam, 1992), for example, Neal Stephenson imagines the wired-and-privileged retreating into autonomous "Burbclaves" complete with customs checkpoints and private security forces. In between: "Lepers roasting dogs on spits over tubs of flaming kerosene. Street people pushing wheelbarrows piled high with dripping clots of million- and billion-dollar-bills that they have raked out of storm sewers. Road kills—enormous road kills—road kills so big that they could only be human beings, smeared out into chunky swaths a block long. Burning roadblocks across major avenues. No franchises anywhere."

19   These will, however, exert a significant influence. See Andrew Gillespie and Kevin Robins, "Geographical Inequalities: The Spatial Bias of the New Communications Technologies," *Journal of Communications* 39, no. 3 (Summer 1989), pp. 7–18.

20   On the diverse appropriations and transformations of telephone technology, see Claude S. Fischer, *America Calling: A Social History of the Telephone to 1940* (Berkeley: University of California Press, 1992).

CHAPTER 6   GETTING TOGETHER

1   Neal Stephenson's *Snow Crash* (New York: Bantam, 1992) did much to pop-
ularize the idea of a virtual meeting place that looked exactly like a physical one,
and was populated by body-double avatars—of varying quality, depending on
what you could afford. His fictional "Metaverse" is organized around the Street, a
brilliantly lit "grand boulevard going all the way around the equator of a black
sphere with a radius of a bit more than ten thousand kilometers." At any time,
millions of people are walking up and down it. On either side there is devel-
opable real estate.

2   On telecommunications, computer networks, and indirect social relationships see Craig Calhoun, "Community without Propinquity Revisited: Categorical Identities, Relational Networks, and Electronic Communication," *Sociological
Inquiry* 68, no. 3 (1998).

3   Michael Dertouzos, presentation to the Club of Rome conference "How
New Media Are Transforming Society," Smithsonian Institution, Washington,
D.C., 1998.

4   Richard S. Tedlow, "Roadkill on the Information Superhighway," *Harvard
Business Review*, November/December 1996. National and international brands,
and the associated marketing strategies, initially flourished in the late nineteenth
century with the rise of high-speed printing, railroads, and efficient mail systems.
Ivory Soap, American Tobacco, Johnson & Johnson, and Coca-Cola were all
founded around 1880. In many ways, the Internet simply continues this story.

5   Manuel Castells, *The Rise of the Network Society* (Malden, Mass.: Blackwell,
1996), p. 364.

6   Quoted in D. J. Czitrom, *Media and the American Mind* (Chapel Hill: University of North Carolina Press, 1982), p. 11.

7   For the classical formulation of the theory of market segmentation, see
Wendell R. Smith, "Product Differentiation and Market Segmentation as Alternative Marketing Strategies," *Journal of Marketing* 21 (July 1956). On the connection to online virtual communities, see John Hagel III and Arthur G. Armstrong,
Net Gain: Expanding Markets through Virtual Communities (Boston: Harvard Business School Press, 1997).

8   For a particularly vivid and moving example, see the story of BostonBill
and the online community of sufferers from the rare disease fibromyalgia. Peter
S. Canellos, "A Champion of the Afflicted Is Mourned," *Boston Globe*, March 16,
1998, pp. A1, A16. On the gay cyberscene, see Michael Joseph Gross, "Good
Thrill Hunting," *Boston Magazine*, April 1998, pp. 50–56.

9   See for example Clifford Stoll, *Silicon Snake Oil: Second Thoughts on the Information Highway* (New York: Anchor, 1996).

10   According to Durkheim, anomie is the condition that results when labor is minutely subdivided, individuals lose sight of the greater purpose of their collective economic efforts, and there is a consequent breakdown of social relations. See Emile Durkheim, *The Division of Labor in Society*, trans. George Simpson (New York: Free Press, 1933; original 1893).

11   Howard Rheingold, *The Virtual Community: Homesteading on the Electronic Frontier* (Reading, Mass.: Addison-Wesley, 1993), p. 2.

12   Stacy Horn, *Cyberville: Clicks, Culture, and the Creation of an Online Town* (New York: Warner Books, 1998), p. 8.

13   Stephen Graham and Simon Marvin, *Telecommunications and the City: Electronic Spaces, Urban Places* (London: Routledge, 1996), pp. 260–263.

14   The Lighthouse of Knowledge (Farol do Saber) system was initiated by Curitiba's mayor Rafael Greca de Macedo. The first was constructed in 1994, and 50 were projected. Its "lighthouse" tower both referred back to the ancient lighthouse and library of Alexandria and evoked a watchtower over the surrounding neighborhood. They are sited near municipal schools and public squares. The mayor, quoted in the Curitiba municipal Web page, says, "The Lighthouses of Knowledge are knowledge terminals, open to the public." They are meant to "chase away darkness and to provide safety to our people, for to know, and to be able to read, are the best safeguard against a world of thieves, illiterates, have-nots, the outcasts of society, excluded from their share of opportunities."

15   The point about dynamic network addressing is subtle but important. Normally, network addresses are associated with particular connection points; that is, for example, how your email gets delivered to the right place. If you want to be able to work at any connection point, without having to log in to a remote machine, you need some simple and effective way of temporarily associating your personal address with that location.

16   Melvin M. Webber, "Order in Diversity: Community without Propinquity," in Lowdon Wingo, ed., *Cities and Space: The Future Use of Urban Land* (Baltimore: Johns Hopkins University Press, 1963), pp. 29–54. See also Melvin M. Webber, "The Urban Place and Nonplace Urban Realm," in Melvin M. Webber, ed., *Explorations into Urban Structure* (Philadelphia: University of Pennsylvania Press, 1964), and Peter Hall, "Revisiting the Nonplace Urban Realm: Have We Come Full Circle?" *International Planning Studies* 1, no. 1 (1996), pp. 7–15.

17   For a cogent introduction to these technologies, their uses, and some of the policy questions that they raise, see Robert B. Gelman with Stanton McCandlish

and Members of the Electronic Frontier Foundation, *Protecting Yourself Online* (San Francisco: HarperEdge, 1998).

18   When I taught at Cambridge University in the 1970s, we did not make much use of electronic telecommunication. Email did not exist, and telephones were few, inefficient, and still thought to be a rather graceless means of communication. If we wanted to meet someone to transact some business, we simply strolled down Trumpington Street at lunch time, or caught up with each other at dinner in college.

19   The point about visibility and invisibility has vigorously been raised by critics of the Habermasian conception of an inclusive, unified public sphere; see Oskar Negt, Alexander Kluge, Peter Labanyi, Owen Daniel, Assenka Oksiloff, and Miriam Hansen, *Public Sphere and Experience: Toward an Analysis of the Bourgeois and Proletarian Public Sphere* (Minneapolis: University of Minnesota Press, 1993). On public space specifically, see Rosalyn Deutsche, "Men in Space," *Artforum*, February 1990, pp. 21–23.

20   Manuel Castells, "Mexico's *Zapatistas*: The First Informational Guerrilla Movement," in *The Power of Identity* (Malden, Mass.: Blackwell, 1997), pp. 72–83. For a prescient RAND Corporation analysis of cyberspace in the revolutions of the future, see John Arquilla and David Ronfeldt, "Cyberwar Is Coming!" *Comparative Strategy* 12, no. 2 (1993), pp. 141–165 (also www.techmgmt.com/restore/cyberwar.htm).

21   Andrew Shapiro, "Internet Treasure," *Boston Review* 23, no. 3–4 (Summer 1998), pp. 18–19.

22   *Democracy in America*, Book 1, chapter XII: "Political Associations in the United States."

23   For a firsthand, practical account of grassroots political organization using online meeting places, see Ed Schwartz, *NetActivism: How Citizens Use the Internet* (Sebastopol, Calif.: O'Reilly, 1996).

24   See Numa Denis Fustel de Coulanges, *The Ancient City: A Study of the Religion, Laws, and Institutions of Greece and Rome* (Baltimore: Johns Hopkins University Press, 1980; original 1864), for a classic discussion of the relationship between *civitas* and *urbs* as traditionally conceived.

25   See, for example, James Dale Davidson and Lord William Rees-Mogg, *The Sovereign Individual* (New York: Simon & Schuster, 1997). For a more nuanced and scholarly treatment, see Saskia Sassen, *Losing Control? Sovereignty in an Age of Globalization* (New York: Columbia University Press, 1996).

26   The more complex current relation of communities and places did not arise from electronic telecommunication, but it has been greatly strengthened by

it. See Claude S. Fischer, *To Dwell among Friends: Personal Networks in Town and City* (Chicago: University of Chicago Press, 1982), and Barry Wellman, "The Community Question," *American Journal of Sociology* 84 (1979), pp. 1201–1231.

27   For more detailed analyses of cities in these terms, from a variety of perspectives, see M. Christine Boyer, *The City of Collective Memory* (Cambridge: MIT Press, 1994); Dolores Hayden, *The Power of Place* (Cambridge: MIT Press, 1995); and Peter G. Rowe, *Civic Realism* (Cambridge: MIT Press, 1997).

CHAPTER 7   REWORKING THE WORKPLACE

1   Walter B. Wriston, *The Twilight of Sovereignty: How the Information Revolution Is Transforming the World* (New York: Scribner's, 1992), p. 61. Wriston provides a vivid firsthand account of the emergence of electronically enabled financial markets.

2   For a survey of the situation in 1998, see "Financial Centers," *The Economist* 347, no. 8067 (May 9, 1998), p. 62. For a good account of the development of electronic trading and an analysis of some of its implications, see Gene I. Rochlin, *Trapped in the Net: The Unanticipated Consequences of Computerization* (Princeton: Princeton University Press, 1997), pp. 74–107.

3   The crash was not only propagated by electronic telecommunications and software; it was also amplified by communications and software failures of various kinds.

4   Matt Richtel, "Record Label to Distribute Music on Line," *New York Times*, May 5, 1999, pp. C1, C9.

5   On automobiles, see Rebecca Morales, *Flexible Production: Restructuring the International Automobile Industry* (Cambridge: Polity, 1994). On clothing, see Edna Bonacich, Lucie Cheng, Norma Chinchilla, Nora Hamilton, and Paul Ong, eds., *Global Production: The Apparel Industry in the Pacific Rim* (Philadelphia: Temple University Press, 1995). On computers, just open up a PC or laptop and take a look at the origin labels on the various components.

6   George Gilder gives many telling examples (you have to filter, of course, for his inimitable up-with-markets, down-with-governments spin) in "The Eclipse of Geopolitics," in *Microcosm: The Quantum Revolution in Economics and Technology* (New York: Simon & Schuster, 1989), pp. 353–370. Of books, he estimates: "A book costs some 80 cents to print; the bulk of its value is created by its author, publisher, distributor, and retailer." Of silicon chips: "Without any physical manifestation at all, the computer design can flow through the global ganglion into another computer attached to a production line anywhere in the world."

7   In the 1980s and 1990s, this became a very fashionable point to make for
popular policy pundits from both the right and the left, each with their own particular spins. George Gilder plugged it in *Microcosm,* for example. And Robert
Reich hammered it in *The Work of Nations: Preparing Ourselves for 21st-Century
Capitalism* (New York: Random House, 1992).

8   For a more detailed account, see Don Tapscott, *The Digital Economy: Promise
and Peril in the Age of Networked Intelligence* (New York: McGraw-Hill, 1996), p.
92.

9   Ithiel de Sola Pool, *Technologies without Boundaries: On Telecommunications in a
Global Age* (Cambridge: Harvard University Press, 1990), pp. 68–69.

10   Ronald H. Coase, "The Nature of the Firm," in *The Firm, the Market, and
the Law* (Chicago: University of Chicago Press, 1990; original 1937), pp. 33–56.

11   See for example Frances Cairncross, "The Future of the Firm," in *The Death
of Distance: How the Communications Revolution Will Change Our Lives* (Boston:
Harvard Business School Press, 1997), pp. 151–153, and Don Tapscott, "Theme
4: Molecularization," in *The Digital Economy: Promise and Peril in the Age of Networked Intelligence* (New York: McGraw-Hill, 1996), pp. 51–54. For a more technical analysis, see Thomas W. Malone, Joanne Yates, and Robert I. Benjamin,
"Electronic Markets and Electronic Hierarchies," *Communications of the ACM* 30,
no. 6 (1987), pp. 484–497.

12   See for example William H. Davidow and Michael S. Malone, *The Virtual
Corporation: Structuring and Revitalizing the Corporation for the 21st Century* (New
York: HarperBusiness, 1993).

13   Lester C. Thurow, "Economic Community and Social Investment," in
Frances Hesselbein, Marshall Goldsmith, Richard Beckhard, and Richard F.
Schubert, eds., *The Community of the Future* (San Francisco: Jossey-Bass Publish-
ers, 1998), p. 25.

14   Since the physical means of production are still often expensive to move,
the threat of withdrawal will often precede or even head off actual withdrawal. It
will pay companies to try to drive down wages and taxes in their current locations to avoid having to pay moving costs.

15   Eric J. Hobsbawm, *Nations and Nationalism since 1780* (Cambridge: Cambridge University Press, 1990), pp. 174–175. Among the best known of these
extraterritorial industrial zones is that of the *maquiladoras* in Mexico's Northern
Industrial Program.

16   Gilder, *Microcosm,* pp. 355–356.

17   What *really* distinguishes Silicon Valley from all the Silicon Wannabees is its unique concentration of specialized human talent, together with the interactions that the concentration allows and the services it supports.

18   Industry leaders whom I interviewed in Bangalore in 1998, for example, consistently mentioned the high-quality local talent pool, attracted by a pleasant environment, good climate, and an established tradition of high-quality education and research institutions, as the key to that city's success in the software export industry. The viability of high-technology campus workplaces on the periphery also depends upon recent investments in roads and on fleets of buses maintained by the larger firms. And real estate development has emphasized plug-and-play, telecommunications-ready buildings that can be occupied and used immediately.

19   Economists typically think of economic communities as structures put into place to define the property rights that are needed to make market economies work, and to defend those rights against internal and external enemies. In the distant past, they frequently corresponded to walled city-states. More recently, they have been coextensive with nation-states. And more recently still, we have seen transnational but still geographic economic communities like the EEC.

20   For an economist's take on these issues, see Thurow, "Economic Community and Social Investment."

21   Diane Coyle, *The Weightless World: Strategies for Managing the Digital Economy* (Cambridge: MIT Press, 1998), p. 210.

CHAPTER 8   THE TELESERVICED CITY

1   Joel E. Tarr, Thomas Finholt, and David Goodman, "The City and the Telegraph: Urban Telecommunications in the Pre-Telephone Era," *Journal of Urban History* 14, no. 1 (November 1987), pp. 38–80.

2   Don Tapscott, *The Digital Economy: Promise and Peril in the Age of Networked Intelligence* (New York: McGraw-Hill, 1995), p. 45.

3   For a cogent, straightforward analysis of the key issues, see Robert B. Gelman and Stanton McCandlish, "Privacy, Anonymity, and Secure Communications: Safeguarding Personal and Business Data in the Information Age," in *Protecting Yourself Online* (San Francisco: HarperEdge, 1998), pp. 35–84. For a more Foucauldian take, see William Bogard, *The Simulation of Surveillance: Hypercontrol in Telematic Societies* (Cambridge: Cambridge University Press, 1996).

4   Collaborative filtering algorithms make use of statistics on consumer choices within a population to extrapolate from past behavior, predict the preferences of

particular members of that population, and thus automatically make personalized recommendations. They are based on the commonsense idea that, if certain individuals have made similar choices in the past, then they probably have similar interests and are likely to make similar choices in the future. When populations are large and choice profiles are lengthy, these algorithms work very well.

5   For a more extended discussion of this development, see Tapscott, *The Digital Economy,* pp. 192–195.

6   Brad Stone and Jennifer Tanaka, "Point, Click and Pay," *Newsweek,* August 17, 1998, pp. 66–67.

7   Bill Gates, "Friction-Free Capitalism," in *The Road Ahead* (New York: Viking, 1995), pp. 157–183.

8   Quoted in Jillian Burt, "Serfing the Net," *21•C,* Spring 1996, p. 69. The Telegarden can be accessed through http://www.usc.edu/dept/garden/.

9   See, for example, Ian W. Hunter, Tilemachos D. Doukoglou, Serge R. Lafontaine, Paul G. Charette, Lynette A. Jones, Mark A. Sagar, Gordon D. Mallinson, and Peter J. Hunter, "A Teleoperated Microsurgical Robot and Associated Virtual Environment for Eye Surgery," *Presence* 2, no. 4 (Fall 1993), pp. 265–280.

10   Paul Krugman, "The Localization of the World Economy," in *Pop Internationalism* (Cambridge: MIT Press, 1997). See also Paul Krugman, *Geography and Trade* (Cambridge: MIT Press, 1993).

11   This is an old and much-reiterated observation, going back at least to Alfred Marshall. The emergence of telecommunications alters the role and effects of this urban glue, but it does not remove it entirely.

12   Paul Krugman, "Technology's Revenge," in *Pop Internationalism,* pp. 191–204.

CHAPTER 9   THE ECONOMY OF PRESENCE

1   Note that *presence* has a variety of related relevant senses. You can be present at a particular place. You can refer to the present time. And you can present yourself.

2   In his late and pedantic *Laws* (737e ff), Plato pinpointed the size of the ideal city-state at 5,040 citizen farmers, plus their families and slaves and some resident aliens. The more empirically inclined Aristotle did not commit himself to a precise number, but specified the relevant boundary conditions. In *Politics* (1326bII) he remarked: "In order to give decisions on matters of justice, and for the pur-

pose of distributing offices on merit, it is necessary that the citizens should know each other and know what kind of people they are." After noting that excessive size also makes it "easy for foreigners and aliens resident in the country to become possessed of citizenship," he went on to summarize: "Here then we have ready to hand the best limits of a state: it must have the largest population consistent with catering for the needs of a self-sufficient life, but not so large that it cannot be easily surveyed. Let that be our way of describing the size of a state."

3   For an insightful analysis of Tiananmen, and the transformations of its role in an era of electronic telecommunications, see Craig Calhoun, "Tiananmen, Television and the Public Sphere: Internationalization of Culture and the Beijing Spring of 1989," *Public Culture* 2, no. 1 (Fall 1989), pp. 54–71.

4   The implications and consequences of the change from oral to literate have, of course, been much discussed; there is a vast literature on the topic. It begins with Plato, who famously argued in *Phaedrus* that is wasn't such a great idea— likely to "create forgetfulness" and "the appearance of wisdom, not true wisdom." In a resonant passage that was endlessly echoed by media critics who followed, particularly McLuhan, Freud wrote, "With every tool man is perfecting his own organs, whether motor or sensory, or is removing limits to their functioning. . . . Writing was in its origin the voice of the absent person." (Sigmund Freud, *Civilization and Its Discontents*, trans. James Strachey [New York: W. W. Norton, 1961; original 1930]). A good modern entry point to the literature is Harold A. Innis's classic *The Bias of Communication* (Toronto: University of Toronto Press, 1951). Another celebrated treatment is Eric A. Havelock, *The Literate Revolution in Greece and Its Cultural Consequences* (Princeton: Princeton University Press, 1982). Marshall McLuhan takes up the topic in chapters 8, 9, and 10 of *Understanding Media: The Extensions of Man* (Cambridge: MIT Press, 1994; original 1964).

5   Lewis Mumford, *The City in History* (London: Secker & Warburg, 1961), p. 97.

6   The earliest portable tablets from Ur recorded simple lists and tallies— devices for keeping track of stored possessions and facilitating the negotiation of exchanges.

7   In *Notre-Dame de Paris* (1831), Victor Hugo dramatized this role of architecture and lamented its apparent passing in an age of printed text. His Archdeacon Frollo famously pronounced, "This will kill that"—architecture would no longer serve as humankind's collective memory. Frollo expanded: "Architecture is becoming ever more tarnished, faded, and dim. The printed word, that cankerworm of the edifice, sucks up and devours architecture, which casts off its raiment and visibly dwindles away. It is shabby, poor, and bare. It no longer expresses anything, not even the memory of another age's art." The historian Anthony Vidler has quipped that France's monumental new national library— detested by many traditional bibliophiles—represents architecture's revenge.

8   Marshall McLuhan, *The Gutenberg Galaxy: The Making of Typographic Man* (London: Routledge & Kegan Paul, 1962), p. 206.

9   For less compressed versions of this long and complicated story, see Warren Chappell, *A Short History of the Printed Word* (New York: Knopf, 1970), and Elizabeth L. Eisenstein, *The Printing Revolution in Early Modern Europe* (Cambridge: Cambridge University Press, 1983).

10   Robert H. Wiebe, preface to *The Search for Order, 1877–1920* (New York: Hill and Wang, 1967).

11   John Dewey, *The Public and Its Problems: An Essay in Political Inquiry* (Chicago: Gateway Books, 1946), pp. 114–115.

12   James R. Beniger, *The Control Revolution: Technological and Economic Origins of the Information Society* (Cambridge: Harvard University Press, 1986).

13   Pioneering work on packet switching was carried out by Paul Baran at the Rand Corporation, and by Donald Davies at the U.K. National Physical Laboratory, in the early and mid-1960s. For the history, see Peter H. Salus, *Casting the Net: From ARPANET to Internet and Beyond* (Reading, Mass.: Addison-Wesley, 1995). For technical details, see Mischa Schwartz, *Telecommunication Networks* (Reading, Mass.: Addison-Wesley, 1987).

14   Packets vary in size, but packets carried by the Internet typically contain about 200 bytes of information—the equivalent of 200 keyboard characters.

15   From the telecommunications engineer's viewpoint, messages are "user" units, and packets are "system" units. The users see messages, but the system handles packets.

16   The routes that packets take may be complex, and packets may pass through dozens of intermediate points between their origins and their destinations. Furthermore, successive packets in the same message may not end up taking the same route.

17   Packet switching thus works best where continuous connection is not required, and where delays can be tolerated—as in much computer-to-computer data exchange, and in transmission of faxes, but not in continuous speech or video communication. The processing overhead created by the need to read and respond to packet addresses is most acceptable where short bursts of data (rather than long, continuous streams) are transmitted. But sufficiently fast packet-switched networks can create the illusion of continuous connection, and thus can sometimes carry synchronous voice and video successfully.

18   In an extensive study published in 1998, researchers at Carnegie Mellon University were able to demonstrate this sort of effect on a Pittsburgh popula-

tion of Internet users—apparently to the great surprise of themselves and many others. See Amy Harmon, "Sad, Lonely World Discovered in Cyberspace," *New York Times,* August 30, 1998, pp. 1, 22. The results are published in Robert Kraut, Michael Patterson, Vicki Lundmark, Sara Kiesler, Tridas Mukophadhyay, and William Scherlis, "Internet Paradox: A Social Technology That Reduces Social Involvement and Psychological Well-Being?" *American Psychologist,* September 1998. They do allow varying interpretation; one commentator remarked, "Are they exposed to the broader world of the Internet, then wonder, 'What am I doing here in Pittsburgh?'"

19 Jennifer Steinhauer, "Old-Line Retailers Resist On-Line Life," *New York Times,* April 20, 1998, pp. D1, D4.

CHAPTER 10   LEAN AND GREEN

1 *Agenda 21* emerged from the Rio de Janeiro environmental summit sponsored by UNCED. See UNCED document A/CONF.151/PC/Add.7, Section 1, Chapter 6.

2 This formulation derives from the World Commission on Environment and Development's definition of "sustainability" as: "Meeting the needs of the present without compromising the ability of future generations to meet their own needs."

3 See, for example, Diane Coyle, *The Weightless World: Strategies for Managing the Digital Economy* (Cambridge: MIT Press, 1998).

4 Lee Goldberg, "The Advent of 'Green' Computer Design," *Computer* 31, no. 9 (September 1998), pp. 16–19.

5 See Pnina Ohanna Plaut, "Telecommunication vs. Transportation," *Access: Research at the University of California Transportation Center,* no. 10 (Spring 1997), pp. 21–26, Ilan Salomon, "Telecommunications and Travel: Substitution or Modified Mobility?" *Journal of Transport Economics and Policy,* September 1985, pp. 219–235, Ilan Salomon, "Telecommunications and Travel Relationships: A Review," *Transportation Research* 20A, no. 3 (1986), pp. 223–238.

6 Peter Hall, *Cities in Civilization* (New York: Pantheon, 1998), p. 960.

7 This strategy has been advocated by Richard Rogers and others. See Richard Rogers, *Cities for a Small Planet* (Boulder: Westview Press, 1997).

8 This idea has been explored in detail by Susan E. Owens. See her *Energy, Planning and Urban Form* (London: Pion, 1986); "Energy, Environmental Sustainability, and Land-Use Planning," in M. J. Breheny, ed., *Sustainable Development and Urban Form* (London: Pion, 1992), pp. 79–105; and "Land-Use Planning for Energy Efficiency," *Applied Energy* 43 (1992), pp. 81–114.

9   This term is not a particularly felicitous one, but it has gained wide currency, so we are probably stuck with it. It has been popularized, in particular, by the
business consultant Stan Davis. See Stan Davis, "Mass Customizing," in *Future
Perfect*, rev. ed. (Reading, Mass.: Addison-Wesley, 1997), pp. 148–197. For further
discussion see B. Joseph Pine, *Mass Customization: The New Frontier in Business
Competition* (Boston: Harvard Business School Press, 1992).

10   Plaut, "Telecommunication vs. Transportation."

ACKNOWLEDGMENTS

This book germinated from the discussions and debates that greeted publication—both on paper and online—of my *City of Bits: Space, Place, and the Infobahn* in 1994. My thanks therefore go to the numerous reviewers, commentators, interviewers, email correspondents, online discussants, students, colleagues, and friends who have raised interesting questions about the interrelationships of cyberspace and urbanism, and who have contributed provocative and insightful ideas to the ongoing discourse. In particular, I want to mention the late Donald Schön, Bish Sanyal, Anne Beamish, Peter Hall, Manuel Castells, Leo Marx, Mel King, and the other participants in the vigorous 1997 MIT colloquium "High Technology and Low-Income Communities"—the proceedings of which have been published as Donald A. Schön, Bish Sanyal, and William J. Mitchell, eds., *High Technology and Low-Income Communities: Prospects for the Positive Use of Advanced Information Technology* (MIT Press, 1998). Kent Larson provided perceptive comments on smart houses, and conversations with Jane Wolfson and Krzysztof Wodiczko illuminated many issues for me. Without this distinguished help, I would have had far less to say.

Finally, this is an appropriate occasion to remember Harvey S. Perloff and Charles W. Moore, who taught me what cities are really for.

INDEX OF NAMES

Rogers, Richard, 80

Royal Flying Doctor Service, 115

Saffo, Paul, 33

Santa Monica, 86

Santarommano, Joseph, 122

Seiko, 55

Siena, 154

Silicon Valley, 81, 144

Singapore, 74, 166n10

Soho (London), 102

Stelarc, 55

Stephenson, Neal, 167n18, 170n1

Stevenson, Robert Louis, 133

Sun Microsystems, 18, 47, 49

Teledesic, 25

Telegarden, 122

Telstra, 25

Thoreau, Henry David, 88

Thurow, Lester, 109

Tönnies, Ferdinand, 22

Tocqueville, Alexis de, 96

Ubiquitous Computing, 60

Venice, 76–77

Venturi, Robert, 34

Videoplace, 39

Webber, Melvin, 75, 93

Weiser, Mark, 60

Well, 23, 90

Wellner, Pierre, 38

Williams, Raymond, 157n1

Wired, 12, 161n28

Wodiczko, Krzysztof, 34

World Wide Web, 12, 13, 17, 25, 28, 48, 92, 118, 119, 127

Wriston, Walter, 100

Xerox Palo Alto Research Center, 35, 60

Yahoo, 122

Zapatistas, 95